MW00573952

From Insight to Innovation

Foreword

Michael Littman

"I was left with my thoughts and imagination," said Jack St. Clair Kilby, describing how he developed the idea for the microchip. The microchip continues to shape our modern world. It allows for the creation of small computer chips with billions of transistors on a small tile of silicon. The microchip is one giant advance that ushered in our modern information age. It is a radical innovation that influences every part of modern daily life. But so is the transistor, a tiny solid-state electronic valve that replaced the vacuum tube. The transistor launched the second age of electronics, and it was the work of two researchers at the Bell Telephone Laboratories, John Bardeen and Walter Brattain. These are but two of the many stories of twentieth-century innovations and insightful engineers and scientists that are the subject of this book.

As described in the preface, this book is the third in a series first conceived of by the author's late father, David Billington, Sr., of Princeton University. The books collectively provide readings for lectures and resources for assignments and term papers about engineers and engineering works that transformed society. The other two books concern engineering innovations and innovators from 1776 to 1883 and 1876 to 1939, respectively. I had the privilege of collaborating and co-teaching with David Sr. for over 20 years. While each chapter about twentieth-century innovation and innovators by David Jr. stands on its own, the book can be very useful as a text for university courses and studies about technology and society.

The topics cover a broad range of twentieth-century engineering works of structures, machines, networks, and processes that transformed our society. Nuclear power for submarines—the disruptive technology advanced by Hyman Rickover, for example—changed the Navy forever and, with it, our national nuclear defense posture and policy. Nuclear-powered submarines submerge almost indefinitely and travel for long distances virtually undetected. This was a global game changer. So, too, are the other technologies or ideas discussed, including the interstate highway system planned under Thomas

MacDonald of the Bureau of Public Roads, and the jet engine that was invented and promoted by Royal Air Force pilot and entrepreneur, Frank Whittle. We can drive coast to coast without traffic lights—this innovation largely replaced the system of national railroads for transport of goods and people. Jet-powered aircraft drastically reduced the time to travel, replacing automobiles and ships for long-distance travel.

Besides stories of transformative technologies of the twentieth century, this well-illustrated book also contains useful material for a general audience about underlying technical ideas without being too technical. And what I like best is the extensive bibliography of primary and secondary materials. Don't read this book without also examining the references—this carefully selected collection of resources is a good place to begin further research about each of these important engineering works and innovators.

From Insight to Innovation

Engineering Ideas That Transformed America in the Twentieth Century

David P. Billington, Jr.

Foreword by Michael Littman

The MIT Press
Cambridge, Massachusetts
London, England

© 2020 Massachusetts Institute of Technology

All rights reserved. No part of this book may be reproduced in any form by any electronic or mechanical means (including photocopying, recording, or information storage and retrieval) without permission in writing from the publisher.

This book was set in ITC Stone Serif Std and ITC Stone Sans Std by New Best-set Typesetters Ltd. Printed and bound in the United States of America.

Library of Congress Cataloging-in-Publication Data

Names: Billington, David P., Jr., 1953- author.
Title: From insight to innovation : engineering ideas that transformed America in the twentieth century / David P. Billington Jr.; foreword by Michael Littman.
Description: Cambridge, Massachusetts : The MIT Press, [2020] | Includes bibliographical references and index.
Identifiers: LCCN 2019057052 | ISBN 9780262044301 (hardcover)
Subjects: LCSH: Engineering—United States—History—20th century. | Technological innovations—United States—History—20th century. | Engineers—20th century.
Classification: LCC TA23 .B483 2020 | DDC 620.00973/09045—dc23
LC record available at https://lccn.loc.gov/2019057052

10 9 8 7 6 5 4 3 2 1

To my late parents,
David and Phyllis Billington,
for a lifetime of love and support

Contents

Preface

Technology is a part of everyday life, yet to most people the engineering that creates it is remote and opaque. Few know much about the principal technologies that underpin modern life, how in a basic sense they work, or how they came about. The aim of this book is to explain, in brief narrative accounts, the origins and basic working of certain key engineering innovations since 1920 that brought America and the world to the early twenty-first century. The book also uniquely describes these innovations through simple numerical formulas or simple graphics. The numbers and concepts in the book are used by engineers but they are presented here without the calculus that professional engineers would need to know.

The United States and many other countries train world-class engineers, but a broader technical literacy in these countries has been an elusive goal. Secondary schools teach natural science and mathematics in ways that emphasize the learning of principles, and these schools and most universities and colleges require nontechnical students to have some exposure to these subjects. But engineering is usually taught to students who plan to become engineers. More recently, at the secondary level, many schools have begun giving students classes in simple engineering design, and science curricula include some engineering ideas. But engineering is usually perceived to be a technical skill or is used as a way to teach science, rather than taught as a set of major innovations that illustrate a kind of insight independent of science.

An informed public should know how key engineering ideas are embodied in the most important objects and systems that are basic to modern life. In research going back to the 1970s, the author's father found that most of the major technical innovations of the last two centuries could be explained in the language of engineers—mathematics—at a numerical level that most American students learn in the first two years of secondary school. In this book, the first three chapters contain boxes that explain engineering ideas in terms of entry-level algebra or geometry. Owing to the greater complexity of the ideas, boxes in the rest of the book explain engineering ideas visually with simple graphics. The main narrative text does not require mathematics to follow.

Engineers also need this kind of overview. Modern engineering knowledge consists mostly of principles and applications that aim to inculcate "best practices," or standards of excellence in design. However, modern engineering is also a sequence of "best works" that began as major innovations and that set new ideals and standards. Scientists, architects, artists, lawyers, and physicians learn the canonical ideas and works in their fields along with their creators. In their training, engineers tend to learn what they know as an abstract body of knowledge in which landmark ideas that are still relevant belong in the form of timeless general principles and applications. In fact, modern engineering also has a body of key ideas exemplified in best works that engineers should know. In addition, this book shows how, in a basic way, engineers can communicate these ideas and works to a wider public.

At their inception, each key innovation usually came from the insight of one or two engineers, and the accounts in this book mostly describe what these people did. In recent decades, historians have highlighted how larger and often neglected social groups contributed to the rise of new technologies. Scholars have also explored the wider context of engineering choices and the social consequences of these choices. Although it describes some of these consequences, and includes several engineers who overcame barriers of race and gender, the present book is not intended to be a broader history. Its main purpose is to provide a basic engineering literacy to a general audience through accounts of innovative engineers, ideas, and works. Learning what these innovators thought and did should be of interest to current and future engineers as well.

The book presents several key innovations in one volume in order to see how they resembled each other and how they differed. A comparison also allows a wider perspective on two issues: the role of science and the role of national government. Perceptions of modern technology tend to treat science and engineering as a single category. While they have much in common, there are important differences between the two that need to be better understood. Public debate over how to stimulate innovation and economic growth has been driven by advocacy in support of a preferred role for government. Although not a study of public authority, the book tries briefly to show what government actually did across a range of key technologies.

This book completes a trilogy on engineering ideas that shaped American civilization over nearly two and a half centuries. In the first volume, the author's father introduced the modern engineering that connected the United States from 1776 to 1883.[1] In the second volume, the author and his father described the engineering breakthroughs that industrialized the country from 1876 to 1939.[2] The first five chapters in the present volume describe the major works of U.S. public engineering in the twentieth century, from great dams and highway programs to the Moon landing in 1969. The remaining three chapters give the main breakthroughs in electronics: the transistor, the microchip, the personal computer, and the Internet. The book combines narrative, technical description, and illustrations to make these insights and works more accessible. Each chapter has a key breakthrough as its focus but also describes some of the more important ancillary innovations.

The title of this book, *From Insight to Innovation*, refers to technical ideas and events that dramatically transformed modern life. Less dramatic improvements are also part of technical development, and the difference between these and the more radical innovations is not always clear. However, other accounts of the most significant twentieth-century innovations include most of those in this book. Two excellent studies by Vaclav Smil give a comprehensive overview of the technical advances of the last two centuries, and an end-of-century volume commissioned by the National Academy of Engineering identifies 20 of the most important breakthroughs of the twentieth century.[3] The present book focuses on the engineering behind a smaller number of these and concludes with an overview and brief look into the future.

Any errors of fact or interpretation are the author's and are not the fault of either the publisher or people consulted in the research and writing of this book. The author welcomes comments sent care of the publisher.

Introduction

The United States of America was an agricultural society when it declared independence in 1776. Modern engineering soon brought change. Robert Fulton used steam engines invented in Britain by James Watt to power the first steamboats, which opened America's inland rivers to settlement and trade in the early nineteenth century. The high-pressure steam engine drove railway locomotives that began to interconnect the country over land from the 1830s onward. The use of water power enabled a textile industry to develop in New England, and steam engines began to power manufacturing later in the century. The electric telegraph invented by Samuel F. B. Morse created a communications network before the Civil War (1861–1865) that spread in the years that followed. At war's end, though, America was still mostly a nation of farmers.

America rose to industrial greatness by the 1920s on the strength of two super-innovations, electricity and internal combustion.[1] These relied on fossil fuels for energy. Before the 1870s, Americans relied on firewood for fuel. Coal had more energy per pound than wood, though, and in the late nineteenth century, coal began to replace wood as a fuel to drive railway locomotives, power factory equipment, and heat homes and workplaces. The coming of motor vehicles after 1890 made petroleum an essential fuel as well, and by the 1920s coal and oil were the nation's primary sources of energy. The environmental drawbacks of fossil fuels would become a concern only later; at the time, they met the nation's need for energy and saved its forests from further destruction.

In the 1880s, Thomas Edison designed a system in New York City to supply electricity to homes and offices from a coal-fueled power plant, and industries to make electric lights, motors, and other devices followed as electric power spread. Alexander Graham Bell used electricity to provide a new form of wireline communication, the telephone; and in the 1920s, Americans began to use electromagnetic waves for wireless radio communication and broadcasting. Television broadcasting over such waves began two decades later.

The modern airplane embodied the principles demonstrated by the Wright brothers in their Flyer of 1903; and motor vehicles became the leading sector of modern industry after 1908, when Henry Ford introduced a mass-produced automobile, the Model T. Airplanes and most automobiles relied on internal combustion engines fueled by gasoline. By the 1930s, new refining processes had increased the amount of gasoline that could be obtained from a barrel of oil, from 10 to 40 percent, and improved its performance in engines. Metallurgical engineering supplied the structural materials needed to make motor vehicles and airplanes, and structural engineers used steel to extend the length of bridges and the height of buildings.

Americans came later to believe that modern industry was the product of modern science. In fact, scientists made their contributions mostly after rather than before the technical breakthroughs outlined above.[2] General Electric and Westinghouse created laboratories in the twentieth century where scientists and engineers developed tungsten light bulbs and alternating current, which were more practical than Edison's original carbon-filament bulbs and direct-current power distribution system. Laboratory engineers streamlined the shapes of automobiles and airplanes in the 1930s, and the Bell Telephone Laboratories advanced knowledge in many fields besides telephone service. But the new industrial research served mostly to improve technologies already in existence. Growth of the resulting industries, along with advances in agriculture and water supply, turned America from a rural into a mostly urban civilization and enabled the nation to achieve victory in World War II (1939–1945).

In some ways, major engineering innovation in the twentieth century did not change. The most important continuity was the extent to which new technologies built on those that went before them. The formula for a reciprocating automobile engine went back to James Watt in the eighteenth century, who used it to describe the action of steam engines, and the principle of rocket propulsion went back to Newton's third law of motion. The first electronic amplifier, the 1906 triode, had to be enclosed by a glass tube with a vacuum inside; the first transistor in 1947 was a triode that could work without a vacuum. When Steve Wozniak designed the Apple II personal computer in 1977, his basic insight was to combine in an innovative way a new kind of microchip, the microprocessor, with a color television monitor. New innovations opened further possibilities, but each also grew from a foundation of skills, ideas, and things that had developed up to the time of the innovation.[3]

The twentieth century differed, though, in requiring most engineers to possess more formal training in order to innovate. Although their formal education was limited, Edison and the Wright brothers were just as brilliant engineers as any in the decades that followed them.[4] The personal computer began in the late 1970s with the founders of Apple and Microsoft, who were self-taught in what they needed to know. But Hoover Dam, the national highway network, nuclear energy, jet airplanes and spacecraft, long-span bridges and tall buildings, and advances in computing required their engineers to have more formal training in engineering and science. Whatever their training, though, all innovators had to design imaginatively and win backing for their ideas.

The role of government also changed. The United States industrialized with the help of a federal government that gave patent protection, levied tariffs to protect industry against imports, and regulated some domestic practices. But before 1920, private industry was able to rise on the strength of private demand. Afterward, huge dams to provide water and power, and new highways to interconnect the country, required public engineering. Following World War II, the federal government grew much larger for reasons of national defense, and new industries in aerospace and electronics depended on military support for their rise and for much of their continued prosperity. However, the role of government varied with each technology. Its role did not prove any blanket view of government as a means, or a hindrance, to innovative insight.

The major innovations after 1920 each tell a unique story. The great dams brought huge rivers under control, opening the western United States to modern life much as steamboats and railroads had opened the middle of the country a century earlier to settlement and trade. The Tennessee Valley Authority was the first attempt to use public engineering to lift a huge region out of poverty. The federal highway program facilitated the change to a largely suburban civilization, while the development of nuclear energy after 1945 brought a new and controversial source of power. With federal backing, the jet engine and the space rocket overcame boundaries of speed, distance, and altitude, although the technologies did not sustain their rapid advance after 1970. The transistor in 1945–1947 was a radical breakthrough in electronics, as was the integrated circuit or microchip in 1958–1959. The electronic computer and the Internet would prove equally radical in their impact. The transistor and microchip originated in the private sector but needed military support to find a market. The military developed early computers and computer networking, but it was the private sector that later gave these innovations a mass market.

After 1920, radical innovators had to succeed in a society dominated by large private and public institutions. Yet the innovations covered in this book showed that unconventional insight was still possible in a more bureaucratic age. What mattered was independent vision, and a society willing and able to respond to it.

The Principles of Modern Engineering

The two volumes that preceded this book provide a framework to understand modern engineering in terms of certain distinctions. These inform the present book and also give an accessible language to describe technical innovation.

Normal and Radical Innovation

Technical innovations are distinguished by their engineering importance to society, and this importance can be considered in three senses. Most of what society considers innovations are improvements to existing goods and services. These advances occur frequently and are often crucial to the success of business firms, but their importance tends to be short-lived. Microprocessors do the principal work inside computers, for example, and for many years microprocessors doubled in capacity roughly every two years. As a result, however, they became obsolete quickly. A more important kind of innovation introduces a major new use for an existing product, or introduces a new product or process that significantly affects an existing industry. The microprocessor was this kind of new product when it first appeared. To the extent that new microprocessors embodied the idea of the original one, the original product continued to be influential. Finally, a small group of innovations have (or have had) a radically transformative effect on society over a long period of time, such as the development over a century ago of electric power and the motor vehicle. The advances that led to today's digital computing may be comparably radical.

The differences between kinds of innovation are not exact, and all are vital to a modern society. The key point is that innovation is not a continuous flow of changes on only one level. For clarity, this book adopts the aeronautical historian Walter Vincenti's distinction of two levels of innovation. Incremental and intermediate innovations can be classified as *normal*, meaning the kinds of innovation that firms and societies do (and need to do) most of the time. The term *radical* describes ideas that transform engineering and society in more basic ways.[5] Admittedly, any choice of these ideas is arbitrary, but most of the events in this book would likely be in any short list of the more radical innovations that occurred in the period from 1920 to the early twenty-first century.

The traditional image of radical innovation is a breakthrough insight by one or two individuals, who then design and demonstrate prototypes. The microchip conceived by Jack Kilby and Robert Noyce came close to this image, although its innovators needed the help of colleagues. Further work remained before microchips were ready for production, and the new chips then had to find a market. However, some radical innovations are better seen as larger group efforts. The U.S. highway program began in the 1920s as the vision of a single leader, Thomas MacDonald, but the Interstate Highway System after 1956 involved a team effort to complete over the next three decades. The rocket owed much of its practicality to the work of Robert Goddard before 1945, but the American space program of the 1960s was too big to be the engineering vision of a single person. Team efforts did not mean, however, that their work was somehow the result of impersonal forces. The work of larger groups was still the work of individuals with the ability to think independently, when necessary, as well as the ability to work together toward a singular end.

Engineering and Science

A second distinction is the relation between the two activities of engineering and science. The two tend to be regarded by the public as a single whole ("science"), usually with an image of science as the source of new insight and engineering as its application. This image is misleading. Scientists and engineers, broadly speaking, each have a core competency. For scientists, it is the *discovery* of new facts about nature, about things that naturally exist. For engineers, it is the *design* of things that do not naturally exist. Until about 1920, most of the radical advances in modern engineering did not rely on science, in the foregoing sense, as a stimulus. Scientific knowledge has been a more important requirement since then, but engineering design insight is still the key to technical innovation.

Scientists and engineers are both creative in the sense that each questions the boundaries of knowledge: scientists challenge our understanding of nature, while engineers challenge our understanding of what we can design. Both groups have also done the other's work as part of their own. But clarity in terms should matter: when either group studies an aspect of nature or a natural property, it is really doing the work of science, and when either engages in design, it is really doing the work of engineering. To use the term "science" to refer to both activities is to obscure the independent insight that is required to design new things.

This insight can be illustrated by a conflict in the 1870s that arose over how to distribute electricity to lamps. Scientists could show that the maximum transfer of electric power in a circuit required the resistance in the lamps to equal the resistance in the power source (resistance is what causes these things to heat). Experts in the 1870s argued that a network to distribute electric power would have to be designed in this way. However, Thomas Edison recognized that using maximum transfer as a basis for design meant losing half of the energy as wasted heat in the power generator. Instead, he designed a generator with low internal resistance, and after a search for the right ones, lamp bulb filaments capable of withstanding high heat. The result was a system that produced enough electricity to meet demand and reduced waste heat from 50 to 10 percent.[6]

The notion that engineering is applied science, in which science supplies a fundamental insight and engineers merely find practical ways to use it, is to miss the kind of independent insight that Edison possessed. Some may call what engineers do applied science, and engineers today often make use of scientific knowledge. But engineers are designers who think independently and are guided by engineering needs.

Four Prototypical Ideas

A third set of concepts helps to organize modern engineering. This book and the two previous volumes divide major breakthroughs into four main types of designed objects or systems: structures, machines, networks, and processes. These categories refer to basic function. A *structure* holds up or holds back weight and works by standing as still as possible. The principal breakthroughs in structure have been large-scale water-control projects, modern highways and bridges, and tall buildings. A *machine* works by moving or by having parts that move; examples have been prime movers such as the stationary steam engine, the railway locomotive, and the internal combustion engines used in automobiles and airplanes. A *network* is a system that transmits something from one place to another with a minimum of loss. The telephone network operates in this way, as do the electric power grid, radio and television broadcasting, and the circuits that make possible modern computing. A *process* is a system that transmutes, or changes one kind of thing into another kind of thing. Processes for turning iron into steel, crude oil into gasoline, and other chemicals into useful products are examples. These four ideas gave rise to the four original branches of modern engineering: civil, mechanical, electrical, and chemical.

Engineering schools today divide themselves into a larger multitude of departments, programs, and specialties, and many engineering ideas involve more than one function: the twentieth century automobile, for example, integrated electricity, chemical combustion, mechanical motion, and a steel framework. But the four ideas of structure, machine, network, and process are helpful to understanding the principles at work in the most important innovations, including those in this book.

Three Perspectives

The last distinction consists of three questions that all engineers must answer before making a design. First, can the object be made and will it be efficient and safe? These questions typically involve physical relationships that must be calculated and tested.

Second, what is the need for, or potential usefulness of, the object? How much will it cost to design and make, and will the benefit be worth the cost? These questions are typically measured in terms of money, which connects engineering to economics and politics. Finally, if the object is practical and useful to make, and will be visible, can it have an appealing design that does not add appreciably to its cost? Will it improve the quality of life and have an acceptable impact on society and on the natural environment? These questions are harder to answer with tests and measurements. Answers rely on the aesthetic vision and ethical judgment of the designer as well as on what the surrounding society wants or will allow.

These concerns imply that the engineer has a freedom to make choices. Any design may be one of several ways to accomplish some purpose, and having this freedom gives engineers room to imagine new possibilities. However, this freedom carries an obligation to make responsible choices: to make efficient use of materials and economical use of public or private funds, to adhere to a high ethical standard, and to enhance the natural environment and human life. Society also has a choice in how it funds and makes use of innovations, but as the designer, the engineer has a responsibility that comes first.

Transformative innovations cannot succeed without financial support from government or private investors, and the contributions of workers and consumers are essential to their realization. Society may or may not be ready for them (ideas can occur to more than one person at nearly the same time) and engineers may or may not anticipate their larger consequences. Such radical innovations have usually begun in the insights of engineers who could see beyond what was generally thought possible. The rest of this book aims to give engineers, interested students, and the general reader a grasp of key technical ideas from the twentieth century and of the engineer-innovators who conceived them.

1 Rivers and Regions: The Age of Frank Crowe

The United States changed after 1865 from an agrarian nation to an increasingly urban and industrial one, and by 1920 a majority of Americans no longer lived on farms.[1] The great private industries in 1920 still included those that were prominent in mid-nineteenth century America: agriculture, textiles, the railways, telegraphy, mining, and the making of iron. To these, the late nineteenth and early twentieth centuries added steel making and steel construction, motor vehicle manufacturing, oil refining and other chemical making, electricity to power lights and other electrical equipment, and the telephone network. Powered aviation and radio transmission were just beginning. Railways and rivers still carried most of the nation's freight, but by the 1920s motor vehicles were replacing horses to move people and goods locally. Motor vehicles required better roads and bridges, and growing cities required new supplies of water and energy.[2]

A public works infrastructure of bridges, roads, and water supplies had grown locally with cities and towns before 1920, as had private utilities to supply gas and later electricity. Public works became larger and more visible after 1920, though, as civil engineers began to reshape larger regions of the United States with paved highways, long-span bridges, and large water-control and electric-power projects. The federal Bureau of Reclamation and the U.S. Army Corps of Engineers built great dams to control flooding, improve inland navigation, and produce electric power. Hoover Dam on the Colorado River in the Southwest was the most radical of these structures, doubling the height of any previous dam. Great dams followed in three other regions west of the Mississippi River: on the rivers of California's Central Valley, on the Columbia River in the Pacific Northwest, and on the Missouri River in the northern Plains states.

Civil engineers working for public authorities, or on contract to them, built the dams to control and extend supplies of water, produce hydroelectric energy, and relieve unemployment. The new dams disrupted the natural ecosystems of the rivers they controlled, and the reservoirs impounded by the dams in arid parts of the country depended on rainfall. At the time, however, these drawbacks were less important to society than the immediate purposes that the dams could serve.

Water and National Growth

Water for drinking and sanitation became a matter of increasing public concern in nineteenth-century America. By the end of the century, cities and towns had begun to make local investments to impound, filter, and carry water from nearby rivers and watersheds to homes and workplaces.[3] In 1919, a public engineer in Maryland, Abel Wolman, confirmed that adding small amounts of chlorine greatly improved the safety of the urban water supply in Baltimore. New York, Chicago, and other cities adopted his recommendations, which became standard practices nationwide.[4]

Water was also a means of transportation. The United States receives abundant rainfall east of the Mississippi River and in the Pacific Northwest. Less falls over the Plains states, and very little falls in the Southwest. The Mississippi River and its two main tributaries, the Missouri and the Ohio, drain the middle of the country between the Appalachian Mountains in the east and the Rocky Mountains in the West (figure 1.1). A number of rivers flow east and west from these two mountain ranges, and the larger rivers were possible to navigate hundreds of miles inland.[5]

After Robert Fulton's successful 1807 steamboat trial on the Hudson River in New York, steamboats traveled up the Mississippi River, opening the middle of the country to settlement and trade. River water also powered textile manufacturing in New England. After 1865, the railroad became the mainstay of American transport, but grain and other bulk goods also continued to move by barge and boat. George Westinghouse pioneered the use of water power to generate electricity in the 1890s, when he installed generators near Niagara Falls and carried electric power 20 miles to Buffalo, New York. The use of hydroelectric power spread.[6]

Two federal agencies had a national influence over water. In 1824, Congress authorized the U.S. Army Corps of Engineers to manage the nation's navigable rivers and waterways. The Corps built levees (raised embankments of earth) alongside the Mississippi River to limit flooding, and the Corps removed natural obstacles in the river, such as sandbars. The Corps also regulated the height of bridges over navigable rivers to assure sufficient clearance for vessels underneath and began in the late nineteenth century to build locks and dams. Canals had long employed locks, compartments that could be filled or drained of water to raise and lower boats, so that the boats could travel up and downhill. In 1885, the Corps began to build a series of "navigation dams" on the Ohio River. Through locks at each dam, boats could be raised or lowered to

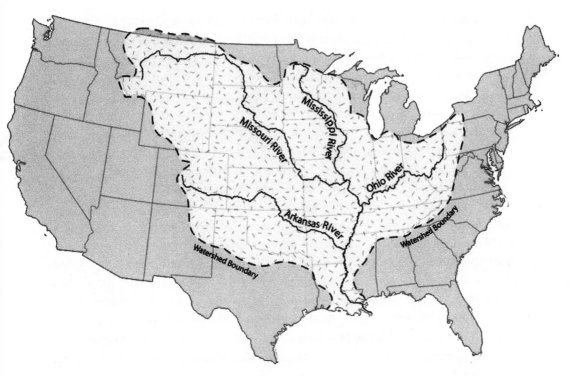

Figure 1.1
The Mississippi River Basin. Courtesy of U.S. National Park Service.

facilitate travel up- and downstream. In the early twentieth century, the Corps began to build navigation dams on the upper Mississippi River as well.[7]

A second federal agency served a different purpose. In 1902, the U.S. Department of the Interior created the Reclamation Service, renamed the Bureau of Reclamation in 1923, to provide irrigation water to small farmers in the states west of the Mississippi. The beneficiaries were to be farmers with 160 acres (64 hectares) or less, who were supposed to repay the cost of building dams to impound the water. The Reclamation Service built a number of irrigation dams in the West but small farmers proved unable to repay the expense. The Bureau struggled to find a role until the federal government provided new funding for larger dams in the 1930s.[8]

The Colorado Compact

A radical leap in twentieth-century dam building began in the American Southwest, an arid region of deserts and gorges cut by a great river, the Colorado. Originating in the mountain state of Colorado, with tributaries there and in Wyoming, New Mexico, Utah, and Arizona, the river flows southwest and divides Nevada and California from Arizona before entering Mexico to empty into the Gulf of California (figure 1.2). The Colorado River descends through a series of deep canyons, including the spectacular Grand Canyon in Arizona. Before the construction of dams along its length, the river discharged water at its mouth at an average rate of 22,000 cubic feet (623 cubic meters, or m^3) per second. This flow reflected the aridity of the region. The Mississippi River discharged water at a rate of almost 600,000 cubic feet (16,990 m^3) per second.[9]

Until the twentieth century, a lack of water limited modern development of the American Southwest. In 1900, a private company began to tap the Colorado River to irrigate nearby desert in southern California. The promoters dug a short canal over the border in Mexico to connect the river to a dry natural channel that crossed back into the United States, where shallower ditches distributed the water to irrigate hundreds of square miles of desert known as the Salton Sink. The promoters renamed the irrigated land on the American side of the border the "Imperial Valley" (figure 1.2) and attracted farmers and a railway line to the outside world. However, during floods from 1904 to 1906, the Colorado River overwhelmed the simple, poorly designed irrigation works and deluged the new farmland. By the time the river was contained in 1907, the lowest part of the area had become a permanent lake, now known as the Salton Sea.[10]

Southern California's growing cities also needed water. The voters of Los Angeles approved funds in 1905 to enable the city's water engineer, William Mulholland, to purchase private farmland in the Owens River Valley to the north, just east of the southern Sierra Nevada Mountains. To draw Owens River water over 233 miles of mountain and desert, Mulholland completed an aqueduct in 1913, increasing the water supply of Los Angeles tenfold. But the city's 1900 population of 102,479 rose to 319,198 in 1910 and to 576,673 in 1920. It would more than double to 1,496,792 by 1930. After 1920, in search of additional water, Los Angeles began scouting possible places to build an aqueduct to the Colorado River.[11]

California's need for water alarmed the other states of the Colorado Basin, whose towns and farmers did not want their huge neighbor to draw all of the river for itself. At Santa Fe, New Mexico, on November 24, 1922, the U.S. Secretary of Commerce, Herbert Hoover, secured the agreement of the seven Colorado Basin states to a compromise known as the Colorado Compact. The states agreed to divide the annual flow of the river, in which the upper basin states of Wyoming, Colorado, Utah, and New Mexico would receive as much water as the lower states of Nevada, Arizona, and California. Later agreements subdivided these shares by state. Arizona withdrew, objecting

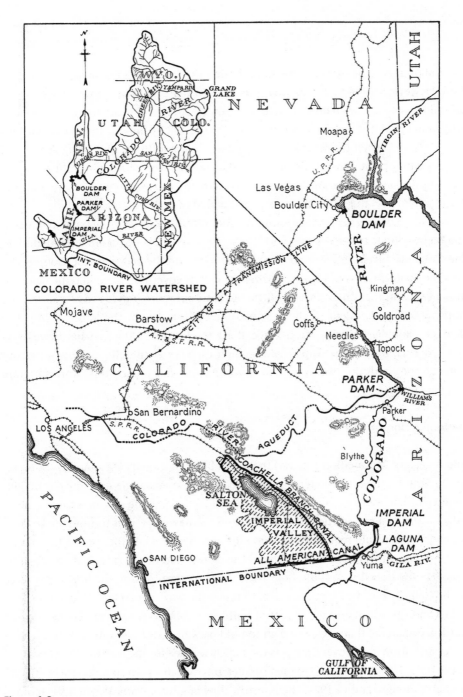

Figure 1.2
Lower Colorado River, also showing the Imperial Valley, in 1941. Inset shows the Colorado Basin.
Hoover Dam was named "Boulder Dam" at the time. Courtesy of U.S. Bureau of Reclamation.

to its allotment of one-third of the lower basin's flow while California received 60 percent. However, in 1963, Arizona finally accepted a share of 38 percent, when Congress agreed to build a new channel system, the Central Arizona Project, to supply the state with water from the Colorado River. As part of the 1922 compact, the federal government, which owned the land along the lower Colorado River, also pledged to build a great dam on the lower Colorado. The dam would control flooding, supply water for irrigation and other uses, and generate electricity, with southern California (as the main beneficiary) to pay for the dam over a period of time.[12]

In 1928, Los Angeles and surrounding communities created a new public agency, the Metropolitan Water District, with the right to levy taxes to build an aqueduct to the Colorado River. The formation of the new District helped persuade Congress to pass the Boulder Canyon Act, authorizing the federal government to build the dam and also a canal on the American side of the Mexican border to provide river water to the Imperial Valley. President Coolidge signed the bill into law in December 1928.[13] The dam site, Boulder Canyon, was a gorge along the Colorado River separating Arizona from Nevada. In fact, the dam would be built on a better site, Black Canyon, twenty miles to the south. In 1930, the Secretary of the Interior named the future structure Hoover Dam.

Designing Hoover Dam

The Bureau of Reclamation received the job of building Hoover Dam, and work began in the early 1920s on the design. The Bureau decided to make the structure out of concrete, but there were two contrasting choices for its design. One was for what engineers called a *gravity dam*, a structure that relied on its mass or weight to hold back water. A gravity dam would require a large amount of concrete. The other possibility was an *arch dam*, in which the form or shape of the dam would be the principal source of its strength. An arched wall could hold back water by transmitting the horizontal force of the water to rock walls on each side, in much the way that an arch bridge transmitted its vertical weight and the weight of traffic to its abutments. An arch dam required much less material than a gravity dam to hold back the same amount of water.[14]

An arch form for Hoover Dam was too radical an idea for the Bureau of Reclamation at the time. The Bureau had designed smaller gravity dams with some curvature to add strength, though, and in 1924, the Bureau's chief engineer, Frank Weymouth, submitted a gravity design for Hoover Dam and gave it an arched curvature (figure 1.3). The curvature wasn't necessary, as the massive weight alone would assure the safety of the

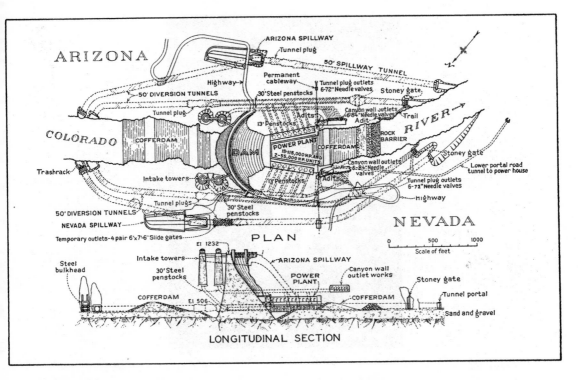

Figure 1.3
Design of Hoover Dam. Upper diagram ("Plan") shows the arched gravity dam in center, with temporary "cofferdams" on either side of the dam site, and with diversion tunnels on either side of the river, to permit construction of the dam on dry ground. Lower diagram ("Longitudinal Section") shows a profile of the dam and cofferdams. Courtesy of U.S. Bureau of Reclamation.

dam. The design was thus conservative; what was radical about the dam would be its unprecedented scale and the speed of its construction. A younger engineer, John L. "Jack" Savage (figure 1.4), took over from Weymouth and refined the design over the next five years.[15]

By the 1920s, civil engineers understood how to design gravity dams, in which the upstream side was nearly vertical and the downstream side had a flatter slope. This shape could be idealized as a right triangle formed by the upstream wall and the base. Engineers also understood the principal forces that a dam needed to resist, chiefly an overturning force exerted by the water pressing on the dam and a smaller upward pressure, called uplift, that could be caused by seepage of water under the dam. With certain formulas (boxes 1.1, 1.2, and 1.3), engineers could determine the

Figure 1.4
John L. Savage. Courtesy of Berkeley-Lainson Studios. Peter Berkeley, photographer.

amount of concrete that a gravity dam would need to resist the water behind it and be safe.[16]

On the night of March 28, 1928, however, a recently completed gravity dam with a slight curvature, the St. Francis Dam, 40 miles northwest of Los Angeles, collapsed in the greatest civil engineering disaster in the United States of the twentieth century. A wall of water roared down a 50-mile long valley and reached the Pacific Ocean, killing more than 400 people. The failure called into question the design for Hoover Dam. However, investigations of the St. Francis disaster blamed defective foundations and other inadequacies in the dam design by its engineer, William Mulholland, and not the choice of an arched gravity dam form. A board of engineers examined the final design for Hoover Dam and still called for more concrete, which would have extended the base of the dam over faults in the bedrock. Bureau engineer Jack Savage was able to resist this change, as the original design already called for close to four times as much concrete as the dam would need to be safe.[17]

Box 1.1
Hoover Dam: Horizontal Force

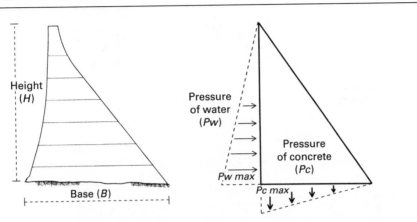

Hoover Dam (profile view) (*left*). Pressures acting on the dam (idealized) (*right*).

A concrete gravity dam has an approximately triangular profile, with the upstream face forming the height of the dam (*H*) at a right angle to the base (*B*).

Water has a density of 62.4 pounds per cubic foot and exerts a maximum pressure at the base of the dam ($P_{W\,max}$) that is equal to the density of the water × the height of the dam in feet. For Hoover Dam, $P_{W\,max}$ = (62.4) (726) = 45,302 pounds per square foot. For simplicity, this can be expressed as 45.3 kips per square foot. One kip, short for 1 kilopound, equals 1,000 pounds.

The total water pressure on the upstream face is known as the horizontal force on the dam (F_H). On a slice of the dam that extends from the top of the dam to its base and is 1 foot wide and 1 foot deep, the horizontal force equals the maximum water pressure at the base × the height of the dam, divided by 2:

$$F_H = (P_{W\,max})\,(H)\,/\,2$$

$$F_H = (45.3)\,(726)\,/\,2 = 16{,}444 \text{ kips}$$

The horizontal force is one of two main forces that engineers must calculate in the design of a dam. The other is the vertical force (see box 1.2).

Source: Dimensions of Hoover Dam are from *Boulder Canyon Project Final Reports*, Part IV—Design and Construction, Bulletin 1, General Features (Denver: U.S. Department of the Interior, Bureau of Reclamation, 1941), p. 137. The formula for horizontal force is taken from David P. Billington and Donald C. Jackson, *Big Dams of the New Deal Era* (Norman: University of Oklahoma Press, 2006), 37.

Box 1.2
Hoover Dam: Vertical Force

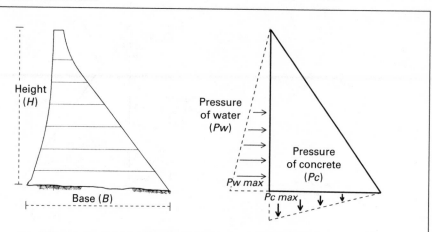

Hoover Dam (profile view) (left). Pressures acting on the dam (right).

The concrete used in Hoover Dam has a density of about 155 pounds per cubic foot and exerts a maximum pressure ($P_{C\,max}$) on the base where it meets the upstream face. At this point, known as the heel of the dam, the vertical pressure (weight of the concrete) equals the density of the concrete × the measurement of the base in feet from the heel to the downstream end, or toe. For Hoover Dam, with a base 660 feet from heel to toe, $P_{C\,max}$ = (155) (660) = 102.3 kips per square foot.

The total pressure of the concrete on the base is called the vertical force in the dam (F_V). On a foot-wide slice, this equals the maximum pressure of the concrete × the base length divided by 2:

$$F_V = (P_{C\,max}) \, (B) \, / \, 2$$

$$F_V = (102.3) \, (660) \, / \, 2 = 33{,}759 \text{ kips}$$

By knowing the horizontal and vertical forces in a dam, the design engineer can calculate a rotational force and a resistance to it that are critical to the dam's safety (see box 1.3).

Source: Density of concrete per cubic foot in Hoover Dam is from *Boulder Canyon Project Final Reports*, Part IV—Design and Construction, Bulletin 1, General Features, 144. Figure given is unit weight, a number that if not a reference to density is very close to it. The formula for vertical force is taken from Billington and Jackson, *Big Dams of the New Deal Era*, 37.

Box 1.3
Hoover Dam: Resistance to Overturning

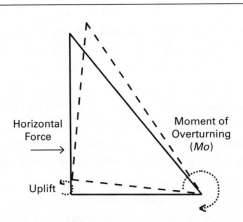

The principal danger to a concrete gravity dam is a rotational overturning force in the downstream direction caused by the horizontal force of the water. A much smaller force in the upstream direction, uplift, can also undermine the base and cause it to slide backward. The downstream rotational force, known as the moment of overturning (M_O), is equal to the horizontal force (F_H) × the height (H) multiplied by 1/3.

In Hoover Dam, F_H = 16,444 kips and H = 726 feet, giving a figure in thousands of foot-pounds, or foot-kips. A foot-pound is a force of 1 pound over a distance of 1 foot in the direction that the force is pushing:

$$M_O = (F_H)\,(H)\,(0.33)$$

$$M_O = (16,444)\,(726)\,(0.33) = 3,939,653 \text{ foot-kips}$$

The dam will resist rotation by an opposite reaction, known as the moment of resistance (M_R), that is equal to the vertical force (F_V) × the base (B) multiplied by 2/3. In Hoover Dam, F_V = 33,759 kips and B = 660 feet:

$$M_R = (F_V)\,(B)\,(0.66)$$

$$M_R = (33,759)\,(660)\,(0.66) = 14,705,420 \text{ foot-kips}$$

By designing the dam with a ratio of M_R / M_O of 3.73, engineers gave Hoover Dam close to four times the resistance that it needed against overturning. This was also more than enough to counteract uplift and any tendency of the base to slide laterally.

Source: Billington and Jackson, *Big Dams of the New Deal Era*, 36–38. Formulas for uplift and sliding are in ibid., 39–41.

Building Hoover Dam

The construction plan for Hoover Dam would require excavating four tunnels inside the canyon walls, two on each side of the Colorado River, with each tunnel roughly three-quarters of a mile long and 50 feet in diameter. These tunnels were to divert the river so that the structure could be erected on a dry river bottom. The dam itself would rise 726 feet (221 meters, or m) above bedrock, twice the height of any previous dam, and at its base would measure 660 feet (201 m) from back to front, tapering (back to front) to 45 feet (13 m) at the top, wide enough to accommodate a paved road. The crest length of the dam, the length of the curved top from one wall of the canyon to the other, would be 1,282 feet (390 m). After the river had been diverted, workers would then build up the dam in concrete.[18] The new engineering of the twentieth century made such a project possible: modern methods of making concrete, electric power to operate cableways and lights, and huge machines powered by internal combustion engines or electricity. When completed, the dam would impound a lake 115 miles (185 kilometers, or km) long and 585 feet (178 m) deep.

After surveying the site to work out detailed specifications, the Bureau of Reclamation solicited bids from private contractors in March 1931. The winning bid of $48.8 million (about $700 million in 2010) came from a consortium of western firms known as the Six Companies, including the dam-building firm of Morrison-Knudsen, the Bechtel engineering company, and Henry Kaiser, a successful industrial entrepreneur. Construction began that summer. To write the winning bid, the Six Companies had turned to a leading construction engineer, Frank Crowe (1882–1946), and afterward hired him to be the engineer in charge of construction for Hoover Dam.[19]

After graduating from the University of Maine in 1905 with a civil engineering degree, Crowe joined the U.S. Reclamation Service and spent almost 20 years building dams throughout the American West, acquiring a reputation for completing projects on time and for less than their budgeted cost. After the renamed Bureau of Reclamation decided to contract its dam-building to private firms in 1924, Crowe took a position with the Idaho-based Morrison-Knudsen company and continued to build dams with a team of skilled workers who followed him from site to site. As the prospect of building Hoover Dam drew closer at the end of the 1920s, Crowe saw the project as within his grasp. "I was wild to build this dam," he later recalled, and his reputation and commitment helped the Six Companies secure the contract.[20] Another Maine graduate and colleague of Crowe's in the Reclamation Service, Walker Young, had stayed with the Bureau after 1924 and served as its engineer at the dam, overseeing the work to make sure that it met the required specifications (figure 1.5).[21]

Figure 1.5
Walker Young (*left*) and Frank Crowe (*right*). Courtesy of U.S. Bureau of Reclamation.

Hoover Dam was the most challenging project of Frank Crowe's career, and the contract called for him to complete it on a rapid timetable. The Six Companies stood to lose $3,000 per day if the river diversion tunnels weren't ready by October 1, 1933, and the whole project had to be finished by March 1938. Crowe needed to supplement his regular crew with new laborers, and work began in the spring of 1931, before proper housing for the workers had been built. Living conditions on the site were primitive, with laborers living in tents or shacks amid daytime temperatures of 120° Fahrenheit (F) (48° Celsius (C)), with bad water and food. After a group of laborers went on strike in protest, Crowe replaced them with others willing to take the jobs in the deepening Depression. But that autumn the dam workers received air-cooled housing, fresh water, ample meals, and medical attention.[22]

During the first year, the principal work was to excavate the four diversion tunnels. Crowe worked his men on the tunnels day and night in three eight-hour shifts. Tunneling required blasting the rock with dynamite, clearing the debris, and blasting again, advancing a few feet at a time. A worker recalled of Crowe that "if he wasn't in his office, he was down at the dam. It'd never surprise me to see him down there at two o'clock in the morning looking around."[23] As the tunnels deepened, exhaust fumes from motor vehicle engines injured some of the workers inside, but the work forged ahead. By early 1933, the tunnels had been dug and lined with concrete (figure 1.6). While workers dug the tunnels deep into the rock, other workers called "high scalers" performed the dangerous work of preparing the canyon walls by drilling or chipping away stone, so as to make a harder and smoother surface to join to the future dam (figure 1.7).[24]

As the site was cleared for the huge dam to rise, the American economy reached a low point. The stock market had crashed in 1929, and by early 1933, the nation's banking system had collapsed, industrial production had fallen by one-half, and one-quarter of the nation's workforce was unemployed. Franklin Roosevelt won election as President of the United States in 1932 on the promise of a "New Deal" for the American people, and he aimed to promote recovery in part by having the federal government build new public works. Following his inauguration in March 1933, Roosevelt affirmed the national importance of the Colorado dam. To avoid giving credit to President Hoover, Roosevelt's Secretary of the Interior changed its name to Boulder Dam.[25]

In the spring of 1933, Frank Crowe raised a temporary barrier, or "cofferdam," upstream to divert the Colorado River into the diversion tunnels, while a cofferdam on the downstream side kept away the backflow. The barriers allowed workers to excavate the river floor down to bedrock. In June, workers began laying down the concrete for the dam itself, mixing concrete from sand, gravel, and cement and then racing to pour

Figure 1.6
Finished diversion tunnel. Courtesy of U.S. Bureau of Reclamation.

it into huge wooden boxes (molds) before it could harden (figure 1.8). The concrete released heat as it set, and pipes ran cold water through each box to accelerate cooling and prevent cracks from forming. Once each block had set, workers removed the wooden formwork and any gaps were sealed with a cement mixture known as grout. The pipes were also filled with grout and sealed.[26]

As the dam rose higher than lifting cranes could reach, Crowe made use of an innovation he had made in building earlier dams, an aerial cableway that could carry huge six-ton buckets of concrete for pouring into the wooden forms below. Beginning carefully at first, Crowe's workers increased their monthly placement of concrete from 28,000 cubic yards (21,407 m³) in June 1933 to 262,000 cubic yards (200,313 m³) in March 1934. When completed in February 1935, 18 months ahead of schedule, the dam contained 3,251,000 cubic yards (2,485,567 m³) of concrete (figure 1.9). President Roosevelt dedicated the dam in a ceremony on September 30, 1935.[27]

Figure 1.7
High scalers atop a cliff. Courtesy of U.S. Bureau of Reclamation.

    The best route for an aqueduct to carry water to Los Angeles led from a place on the Colorado River to the south, near Parker, Arizona. Here in 1938 the Metropolitan Water District of Southern California completed a smaller structure, Parker Dam, and a 242-mile aqueduct linking the dam to the Los Angeles area. Hoover Dam supplied the electricity needed to pump water from the reservoir behind Parker Dam over mountains and flatlands to Los Angeles. Farther to the south, a low dam at Yuma, Arizona, fed water to a new canal that Bureau engineers built on the American side of the Mexican border. The new canal linked the Colorado River to the Imperial and nearby Coachella Valleys. With these dams in place, agriculture could resume safely in the southern California desert. In 1938, the powerhouses at Hoover Dam also began sending electric power to Los Angeles over newly built transmission lines, and water carried by the Colorado Aqueduct to the city began to flow in 1941. In 1947, a Republican Congress changed the name of Boulder Dam back to Hoover Dam, its designation today.[28]

Figure 1.8
Concrete placement at Hoover Dam. Courtesy of U.S. Bureau of Reclamation.

Although not radically innovative in its overall design, Hoover Dam marked a radical achievement in its construction and unprecedented height. However, other projects, in California's Central Valley, the Pacific Northwest, and the Missouri River Basin, challenged civil engineers in different ways. Together they restructured the American landscape west of the Mississippi River.

The Central Valley Project

The Central Valley, 450 miles (724 km) long, follows two rivers down the length of inland California. The Sacramento River begins in mountains near the border with Oregon and flows to a delta south of the state capital, Sacramento, where it meets the San Joaquin River, flowing north from the Sierra Nevada Mountains near Fresno. The

Figure 1.9
Hoover Dam in 1941. Courtesy of U.S. National Archives. Ansel Adams, photographer.

two rivers then empty into San Francisco Bay. Two-thirds of the rain in the Central Valley falls in the northern part of the valley and often causes the Sacramento River to flood, while the San Joaquin River and southern valley receive little rain. In 1923, state engineers issued a report urging the construction of reservoirs and channels in the north to control flooding and bring excess water to the south. By the early 1930s, these ideas had evolved into a more comprehensive plan to build a dam near the sources of each river, irrigate the northern and southern valleys, and generate electric power. In 1937, the Roosevelt administration assigned the Bureau of Reclamation to build and operate the two dams as part of a Central Valley Project (figure 1.10).[29]

The northern dam became known as Shasta Dam, for nearby Mt. Shasta, and would impound the headwaters of the Sacramento River. Frank Crowe agreed to build Shasta Dam for the engineering firm Pacific Constructors, which won the contract. The Shasta design, also by Jack Savage, was another huge concrete gravity structure, this one with

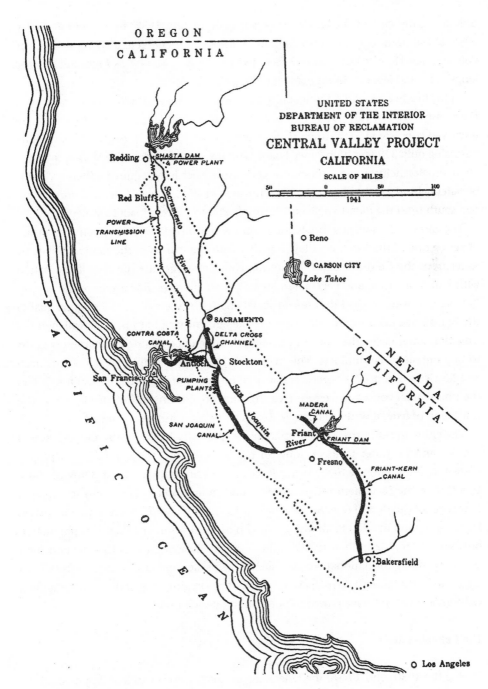

Figure 1.10
Map of the Central Valley Project. Courtesy of U.S. Bureau of Reclamation.

only a slightly curved shape. Crowe began work in July 1938. By its completion in 1945, Shasta Dam contained 6.5 million cubic yards of concrete, making it larger by volume than Hoover Dam. At 602 feet (183 m), Shasta was not as high, but its crest length of 3,460 feet (1,054 m) made it twice as long.

At the headwaters of the San Joaquin River, work on the southern dam, named Friant Dam, also began in July 1938. Two contractors from Los Angeles built the dam as an uncurved concrete gravity structure, with almost the same crest length as Shasta Dam but rising only 319 feet high, since Friant Dam was not intended, like Shasta, to generate hydroelectricity (to be economical, generating electricity required a higher volume of water behind a dam). Friant Dam was finished in 1942. Canals extended north, west, and south from the dam to deliver irrigation water to the San Joaquin Valley.[30]

The dams and associated canals and pumping stations allowed a dramatic expansion of agriculture in the Central Valley. Together with the valleys in the south supplied by water from the Colorado River, California could now provide the rest of the nation with half of its fruits and vegetables, including for the first time a year-round supply. But the new water caused unforeseen social change in the valley. The original goal of the Bureau had been to provide water to independent small farms. The availability of irrigation water raised the value of the land, though, encouraging large-farm operators to buy out small landholders. The owners of large farms employed poorer Mexican and Mexican American laborers, and as a result, the water projects in California had the unintended consequence of increasing rural poverty instead of alleviating it. The restriction of federal water to farms of 160 acres or less was not enforced.[31]

The cities and suburbs of California grew explosively in the 1940s and 1950s, and the demand for water to meet nonfarm needs increased. In the 1960s, in a project known as the California State Water Project, the state built a dam at Oroville, on a tributary of the Sacramento River, to impound more water in the north to supply a new series of canals and reservoirs to the south. The population of California increased from 3.4 million in 1920 to almost 20 million in 1970, while the total U.S. population only grew from 102 million to 203 million people. New industries that depended on federal spending for national defense, such as aerospace and electronics, contributed a large part of this growth after 1945. But without the engineering of new water supplies, California could not have grown in population by such numbers.[32]

The Columbia River

During the 1930s and 1940s, the federal government also began building dams along the Columbia River in the Pacific Northwest. The Columbia flows from the province of British Columbia in Canada south into Washington State. A number of tributaries, chiefly the Snake River from the east, join the Columbia, which then flows west to form part of the state border with Oregon before emptying into the Pacific Ocean (figure 1.11). Unlike the Colorado, the Columbia flows through a region of heavy rainfall, discharging an average of 265,000 cubic feet (7,504 m³) per second at its mouth, more than 10 times as much water as the Colorado River before Hoover Dam.[33]

To tap this water and its power, the U.S. Army Corps of Engineers built a dam on the Columbia River at Bonneville, 40 miles east of Portland, Oregon, while the Bureau of Reclamation built a dam farther upstream at Grand Coulee in eastern Washington. To relieve unemployment in the 1930s, the U.S. Senators from each state agreed to support a dam in the other state if their own state received one as well (Oregon wanted to build the dam at Bonneville). The Roosevelt administration initially budgeted Grand Coulee to be a low irrigation-only dam, but engineering opinion attacked the cost as not worth the benefit it would bring, and Congress found the money in 1935 to build a higher dam at Grand Coulee that could also produce electric power.[34]

Started in 1933, Bonneville Dam was a concrete gravity dam between two islands in the river, with side sections to connect each island to the river banks. The designer of the main "spillway" section in the middle was George Gerdes, a civilian who worked for the Corps of Engineers. Instead of blocking the Columbia to create a lake, Gerdes designed the spillway structure at Bonneville as a low "run of the river" dam with a row of gates having a crest length of 1,450 feet (483 m). During the spring, opening the gates would allow floodwater to pour over the dam, while through the rest of the year, the gates could be closed to keep the water level behind the structure high enough to power electric generators. The dam would also cause the year-round inundation of the Cascade Rapids upstream, allowing oceangoing ships to travel upriver after going through navigation locks next to the dam itself. After clearing the central site for the spillway structure inside cofferdams, army engineers cut lateral notches across the bedrock to hold the concrete base of the dam against sliding, and concrete surfaces and baffles on the downstream side prevented the overflow of water from scouring the base. Once the main spillway dam was complete, the side sections connected the islands to the shorelines. The project finished in 1937 (figure 1.12).[35]

The final design for Grand Coulee called for a high concrete gravity dam to impound water for irrigation and electric power. The Bureau set the height of the dam at 550 feet (167 m) above bedrock and gave it a base, front to back, of 450 feet (137 m). The crest length was 3,867 feet (1,178 m), with a wing wall that increased the crest length to 5,223 feet (1,591 m). Jack Savage designed the dam and made the main wall a massive gravity dam without any curvature. Unlike Hoover Dam, where

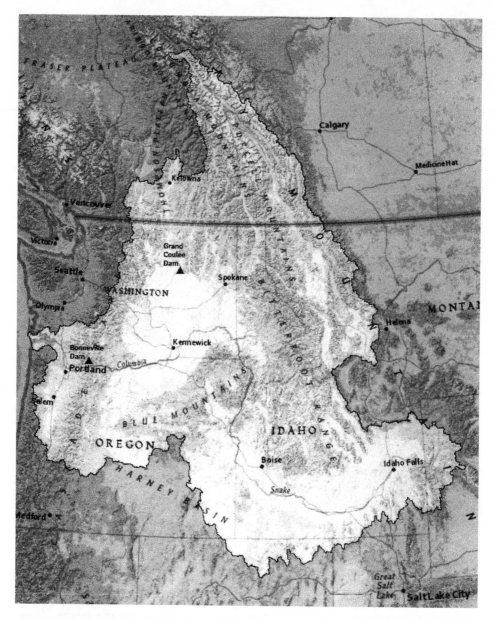

Figure 1.11
The Columbia River Basin, showing Bonneville and Grand Coulee Dams (as triangles). Courtesy
of U.S. Environmental Protection Agency.

Figure 1.12
Bonneville Dam. Main spillway dam with fish ladder in foreground. Courtesy of U.S. Library of Congress.

Frank Crowe built the structure with a Bureau engineer, Walker Young, to observe, the Bureau engineer overseeing Grand Coulee, Frank Banks, also took charge of construction. Grand Coulee needed 12 million cubic yards of concrete, enough to contain the three pyramids of Giza, Egypt. Work began in 1933 and finished in 1941 (figure 1.13).[36]

Bonneville and Grand Coulee Dams finished in time to make an unplanned but vital contribution to American production during World War II. Airplanes had to be made of aluminum, a lighter metal than steel that required enormous amounts of electricity to refine from aluminum ore. To smelt the ore, this electricity had to take the form of direct current, which (unlike the alternating current used in most homes and workplaces) could not travel over long distances. As a result, aluminum makers

Figure 1.13
Grand Coulee Dam. Courtesy of U.S. Library of Congress. Jet Lowe, photographer.

located smelting works near the dams in the Pacific Northwest and near dams else-where around the country that could supply hydroelectric power nearby.[37]

Bonneville and Grand Coulee could not control flooding on the lower Columbia River, though. After an inundation of Portland, Oregon, in 1948, Colonel William Whipple of the Army Corps of Engineers drafted a master plan to control flooding through a series of new dams on the Columbia and its tributaries. Largely completed in the 1970s, the new dams brought the Columbia River under greater control, added electrical generating capacity, and made navigation by oceangoing ships on the Columbia and Snake Rivers practical as far as Lewiston, Idaho.[38]

During the 1950s and 1960s, the dams at Bonneville and Grand Coulee supplied the Pacific Northwest with inexpensive electric power through a new federal agency, the Bonneville Power Administration (BPA), created in 1937, which sold power to public and private utilities in the region. Washington and Oregon nearly tripled in population from 1920 to 1970, while the population of the United States as a whole only doubled. Hydroelectric power helped make this regional growth possible.[39]

The benefits of dams in the Pacific Northwest came at a cost, though, to river ecology. Army engineers recognized in the 1930s that dams on the Columbia River would disrupt the migration of salmon that spawned upstream. At Bonneville, the Corps of Engineers built fish ladders, stepped pools to give fish migrating up and down the river a way around the dam. These efforts failed to prevent the salmon population from undergoing a massive decline. Dams were not the only cause of this collapse but were the main reason for it. In recent years, federal agencies have improved the survival rate of migrating fish, but preservation continues to be in tension with the region's need for electrical energy and water control that the dams provide.[40]

The Missouri River

After the Ohio, the largest tributary of the Mississippi was the Missouri River, which began in Montana and joined the Mississippi near St. Louis. The flow of the Missouri alternated between extremes of very low water that made barge navigation difficult and flooding that made settlement along its length hazardous. In 1933, the Army Corps of Engineers began building a dam at Fort Peck, Montana, in response (figure 1.14). With the ground made of earth rather than rock, the Corps decided to make the dam mostly out of earth because adhesion would be better. Unlike a concrete gravity dam, though, an earth embankment dam needs a much larger volume of material.

The Fort Peck design called for an embankment dam 2 miles long made of 126 million cubic yards of earth, with a steel sheet wall inside to prevent seepage. A spillway alongside would allow the river passage around the dam and would permit regulation of the water level downstream. The dam design of 250 feet (76 m) high called for a base that would measure 4 feet, front to back, for every rise of 1 foot. The challenge to raising such a dam was to move to the site 10 times the volume of material used to build the Grand Coulee Dam. The Corps decided to rely on a recently devised construction method, hydraulic fill, in which a slurry of earth and water would be dredged from the river bottom and piped to the construction site (figure 1.15). Soil mechanics engineers believed that the mud would dry and be suitable for an earth dam. Work on river diversion tunnels (dug into soft shale) finished quickly in 1934, allowing the dam site to be cleared with cofferdams and the inner sheet wall to go up. The Corps then dredged the river floor upstream and piped the mud with water to the dam site. Alongside the dam, the Corps also built a concrete spillway and facilities to generate electricity.[41]

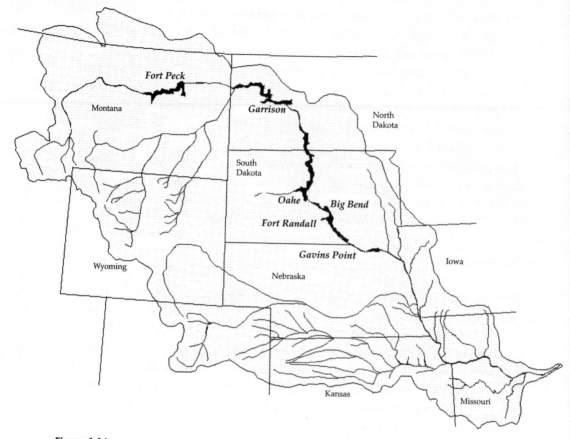

Figure 1.14
Map of the Missouri River Basin marking the six "mainstem" dams on the river today (Fort Peck, Garrison, Oahe, Big Bend, Fort Randall, and Gavins Point). Courtesy of U.S. Army Corps of Engineers.

The project was nearing completion when, on the morning of September 22, 1938, workers noticed that surface water at the southern end of the dam was not draining properly. Then at 1:20 PM, a surveyor wrote: "I was having trouble setting up . . . when I noticed a small crack in the ground underneath my feet . . . in a few seconds it got considerably deeper and longer."[42] The southern end of the dam then gave way in an enormous slide (figures 1.16 and 1.17). The surveyor survived but several workers died. Afterward, engineers redesigned the dam with a slope to rise 1 foot for every 20 feet of distance from front to back. Instead of hydraulic fill, the Corps finished the dam two

Figure 1.15
Using pumped hydraulic fill to raise the Fort Peck Dam. Courtesy of U.S. Army Corps of Engineers.

years later with dry earth brought by truck and compacted with heavy motorized rollers (figure 1.18).[43]

During World War II, the Corps of Engineers and the Bureau of Reclamation decided to build additional dams on the Missouri River downstream. Colonel Lewis Pick of the Corps made plans to construct a few large dams to control flooding and generate electricity, while the Bureau under William Sloan planned many smaller dams on tributaries to facilitate irrigation. Although rivals, the two agencies agreed to support each other's plans in 1944 to prevent the creation of an independent federal authority to control the river. After the war, the Corps built Garrison Dam in North Dakota, Fort Randall Dam in South Dakota, and three more large "mainstem" dams on the Missouri River, while the Bureau built a series of smaller projects on tributaries. Army engineers created the large dams out of dry earth compacted by rollers.[44]

Figure 1.16
Fort Peck Dam, facing south, before the slide of September 22, 1938. Courtesy of U.S. Army Corps
of Engineers and Robert A. Midthun, Jr. Robert A. Midthun, photographer.

The Missouri River caused controversy even with the new dams. In periods of low
flow, farmers wanted the water to be impounded by the dams to irrigate crops, while
owners of barges and boats wanted the water released into the river to provide sufficient water depth so that grain and other bulk goods could move downstream. In years
of heavy rain or snowmelt, dam managers had to release reservoir water to prevent
overtopping, bringing some flooding to communities downstream. But overall the Missouri became less dangerous to live beside, as the large earth dams limited flooding and
brought more regular supplies of water for river transport and irrigation.

Unlike the states along the Pacific Coast, the population of the Missouri Basin did
not grow as quickly, increasing only 30 percent, from 13 to 17 million, between 1920
and 1970. Most of the agriculture on the Plains was grain farming, which was easier to
mechanize than fruit and vegetable production in California. The region also did not
benefit as much from federal defense spending during and after World War II.[45]

Figure 1.17
Fort Peck Dam after the slide. Courtesy of U.S. Army Corps of Engineers and Robert A. Midthun, Jr. Robert A. Midthun, photographer.

Figure 1.18
Fort Peck Dam today, facing north. Courtesy of U.S. Army Corps of Engineers. Robert Etzel, photographer.

Legacy and Future

Although constructed in the 1930s, the great dams originated in the 1920s. Planning for Hoover Dam began after the Colorado Compact of 1922, and the Central Valley Project had its roots in a state initiative of 1923. The dams on the Columbia and Missouri Rivers had their origins in the 1925 Rivers and Harbors Act, when Congress ordered the Corps of Engineers to survey major rivers in the United States to assess their potential to produce hydroelectricity. The surveys, known as the "308 Reports," provided initial studies for the Columbia and Missouri dams.[46]

In the 1920s, though, the federal government did not plan itself to build dams and electric power-generating facilities alongside them; Congress and federal agencies assumed that private utilities would make these investments if they were worthwhile to make. However, the Depression of the early 1930s made private spending on such a scale impossible, and in 1933 the federal government began the ambitious projects that followed Hoover Dam, partly to relieve unemployment and partly to build for the public infrastructure needs of a less depressed American future.

The greatest dam builder of the twentieth century, Frank Crowe, was an emblematic figure of this era in public works. Although the Bureau and the Corps made some innovations, the great dams overall were conservatively designed. The construction engineers were the innovators, with Crowe demonstrating how to excavate and build on an unprecedented scale with extraordinary speed. Although dam construction did not employ more than a fraction of the people rendered jobless by the Depression, the projects gave purpose to much of the nation at a low point in its history.

The dams were not without controversy at the time. Private utilities accepted federal construction of the western dams when it became clear that the utilities could buy and resell the electric power to customers. Criticism came from those who opposed any extension of federal activity and from those who did not see an economic need for such expenditures during a depression.[47] More tragically, several of the dams submerged ancestral lands of Native Americans, including the Wintu behind Shasta Dam; the Mandan, Hidatsa, and Arikara behind Garrison Dam; and the Colville Nation behind Grand Coulee. The tribes protested the new projects, to no avail.[48]

Hydropower supplied one-third of the nation's electricity by 1940 and added critically needed electric power for wartime use after 1941. Demand for electricity after World War II rose faster than existing dams could supply it, though, and the nation reached a limit to the number of places where new large dams could be built. During the late twentieth century, the share that hydroelectricity contributed (in relative terms) declined to its present level of about 9 percent of U.S. electric power generated.[49]

The dams in the American Southwest depend on having sufficient rainfall to fill the reservoirs they impound. By the end of the twentieth century, evaporation prolonged by drought had reduced water levels behind Hoover and other dams. In early 2017, the Oroville Dam in California nearly failed when the opposite problem, unusual rainfall, required use of an emergency spillway that was not adequate.[50] Over the next several centuries, silt buildup behind the dams will eventually require them to come down. Lake Mead, the reservoir behind Hoover Dam, in fact began filling up with silt faster than expected.[51]

To reduce silt buildup and give the upper Colorado Basin states more stored water for use in periods of drought, the Bureau of Reclamation decided in 1956 to build two new dams in Utah on a tributary of the Colorado, the Green River. The dams would have submerged Dinosaur National Monument, though, and environmental conservationists led by the Sierra Club objected. In a compromise, the Bureau agreed to cancel the dams, and the Sierra Club (in a decision that it later regretted) agreed not to challenge the building of a new dam by the Bureau on the Colorado River at Glen Canyon in Arizona, a more remote gorge. The Glen Canyon dam was closer than Hoover Dam to a true arch design, as high as the latter but with a base covering only half the area. Completed in 1966, the dam became part of the Colorado River Storage Project, which provided water to the upper basin states. However, if rainfall in the Southwest is inadequate, Glen Canyon Dam may be taken down to give Lake Mead enough water to supply the larger population that depends on Hoover Dam.[52]

During the 1930s, a third federal agency, the Tennessee Valley Authority (TVA), employed dams and other measures to create a more modern and prosperous life in a water-rich but depressed region of the East. Through great water and power projects, the federal government facilitated postwar growth west of the Mississippi. The leaders of the TVA tried in a more direct way to use public engineering to remake society itself.

The eastern United States has more abundant rainfall than the states west of the Mississippi River, and there was less federal engineering of great eastern rivers than of great western ones. In 1933, however, President Roosevelt and Congress began an ambitious program to restructure the valley of the Tennessee River. The Tennessee Valley was a poverty-stricken rural area in the eastern United States, and a new federal agency, the Tennessee Valley Authority (TVA), received a mandate to control flooding of the river and make navigation on it easier, improve land use practices in the valley, and generate and distribute hydroelectric power to its residents. The President wanted the TVA to set an example of how to use public engineering, not just to supply the physical needs of a region for water and power but to improve its way of life.

The TVA began with three directors. Arthur Morgan, a civil engineer, supervised the building of dams on the Tennessee River and several of its tributaries. An expert on farming and forestry, Harcourt Morgan (no relation to Arthur) instituted soil and forest conservation; and an attorney and former utility regulator, David Lilienthal, created a network to distribute electric power at rates affordable to the region's people. However, two of the directors clashed over the larger social vision of the agency. Arthur Morgan wanted the Tennessee Valley to be a society of small and self-sufficient communities, while David Lilienthal wanted the TVA to be a lever to national economic recovery and an example of regional government. Lilienthal dominated the TVA after Arthur Morgan's forced departure in 1938. However, World War II intervened, and afterward, the rest of the country chose not to set up similar regional authorities. The TVA continued to supply the Tennessee Valley with electricity after the war, and to keep up with postwar demand the agency built new power plants in the 1950s and 1960s that used coal and nuclear energy to generate electric power. Nuclear plants were controversial, though, as were coal-mining practices that came into conflict with the aim of land restoration that had guided the early TVA. Its founders had tried in different ways to create a model of a better society. Their successors are now looking to find a balance between regional demand for electricity and the need for sustainable ways to produce it.

Arthur Morgan and The Miami Conservancy District

Arthur Ernest Morgan (1878–1975) was born in Cincinnati and grew up in Minnesota in a family of modest means. Growing up lonely on account of poor health, he immersed himself in books and ideas, and he came to believe in the need for progressive social reforms. He also held himself and those around him to a high standard of ethical conduct, which would give his approach to engineering its distinctiveness but would lead to difficulties in later life when clashes with other reformers led him to question their ethics.[1]

Morgan could not afford a university education but was able to apprentice himself to civil engineers. By 1910, he was a leading expert on drainage, the removal of unwanted water from land, a specialty that did not require him to have earned a formal degree.[2] In March 1913, a tributary of the Ohio River, the Miami (known now as the Great Miami River), flooded cities and towns along its length in southwestern Ohio, including the city of Dayton (figure 2.1). Local leaders hired Arthur Morgan to advise them on how to prevent a future disaster. He backed the decision of the Ohio state legislature to authorize public "conservancy districts" around the state to manage rivers and control flooding. The Miami Conservancy District was formed in 1915 to prevent future flooding of the Miami River, and Morgan became its chief engineer (figure 2.2).[3]

The standard way to control flooding was to build raised embankments or levees alongside rivers. In 1851, Congress had commissioned two studies on how to limit flooding of the Mississippi River below Cape Girardeau, Missouri. The first study, by Charles Ellet, a private civil engineer, called for building levees, but Ellet also urged the creation of outlets and basins alongside the river to divert floodwater and then release it back to the river gradually. Ellet argued that levees would cause silt to build up on the riverbed and eventually raise the bed; outlets and basins would reduce this danger.[4] The other study, by Captain Andrew Humphreys of the U.S. Army Corps of Engineers, came to the opposite conclusion. Humphreys argued that Ellet's proposal was not worth its cost and that confining the river between levees would accelerate the flow of water and scour the river bottom, preventing the buildup of silt. Also, settlers alongside the Mississippi did not want to remove large amounts of land for water catchment basins. After the Civil War, the Corps of Engineers, responsible for flood control, adopted a "levees only" policy along the Mississippi. But the riverbed in fact

Figure 2.1
The 1913 Miami River flood in Dayton, Ohio. Courtesy of The Miami Conservancy District.

rose with each passing decade, just as Ellet had predicted, requiring levees to become larger and higher to contain the river's floodwaters.[5]

Morgan agreed that levees alone were not the answer to flood control. For the Miami Valley, he proposed to build earth dams across several tributaries to enclose basins on dry land. Permanent openings, or conduits, in each dam would allow the tributary to flow through the dam at a normal rate, but during storms the dam would hold back floodwater, slowing the outflow to a rate that would be safer downstream. The Miami Conservancy District accepted Morgan's proposal.[6] After litigation settled by state courts in its favor, the District acquired some prime land for the dams and basins, removing this land from residential and industrial use. Morgan then gave the District an engineering organization, hiring 2,000 workers. From 1918 to 1922, these workers erected the system of earth dams and basins on the tributaries of the Miami River, with a few levees along the river itself (figure 2.3). Dams with permanent openings were an innovation and Morgan embraced them, in keeping with his view that "every possibility . . . should be explored, with effort to become aware of unrealized and unexpected ways of approach."[7] In later rainstorms, the dams and basins caught

Figure 2.2
Arthur E. Morgan. Courtesy of Antiochiana, Olive Kettering Library, Antioch College.

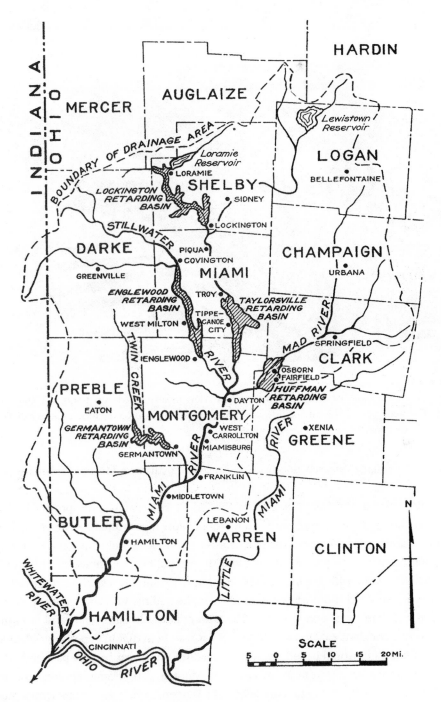

Figure 2.3
The Miami River and tributaries. Shaded areas show basins where floodwaters could be held temporarily and gradually released. Courtesy of The Miami Conservancy District.

© 1913–2010 Miami Conservancy District — All rights reserved

Figure 2.4
Germantown Dam during construction; upstream view into finished outlet showing concrete steps. The steps slowed the rate of flow and reduced erosion to the concrete. Courtesy of The Miami Conservancy District.

and then gradually released floodwaters, preventing catastrophic floods (figures 2.4 and 2.5).[8]

While it lasted, the Miami work was the largest civil engineering project in the country, and it attracted applicants from whom Morgan recruited a staff of capable engineers and workers. To avoid the shack housing and social problems associated with temporary construction projects at the time, Morgan built cottage houses for his workers, including those with families, in several camps. To embody a spirit of cooperation between workers and those who employed them, he had his workers elect councils to manage their temporary communities. The District also sold back some land to farmers, reserving the right to flood it if necessary. Morgan held that in depositing valuable silt, the floodwater would benefit farming naturally. Other areas became parkland.[9]

During and after his work as an engineer, Morgan also became an educator. In 1917, he founded the Moraine Park School in Dayton, which gave high school students considerable freedom to design their own studies and obtain work experience. In 1921, Morgan became president of Antioch College in nearby Yellow Springs, where

Figure 2.5
Germantown Dam; view of outlet, showing discharge during flood. Courtesy of The Miami Conservancy District.

he introduced a curriculum that mixed study with outside work and stressed independent study. He also attached a primary school where children learned responsibility by sharing school governance with teachers. As an educator, Morgan's aim was to train students to be self-reliant and enterprising but also to serve society and uphold a high ethical standard.[10] His work as an educator brought him to the attention of Eleanor Roosevelt in the 1920s. Morgan's engineering experience and idealism appealed to her husband Franklin when in 1933 he became President of the United States and began looking for an engineer to direct the new Tennessee Valley Authority.[11]

The Tennessee River and Wilson Dam

The Tennessee River began in the Appalachian Mountains of eastern Tennessee and flowed west, dipping into northern Alabama and then going up to Paducah, Kentucky, where it joined the Ohio River (figure 2.6). Just before reentering Tennessee from Alabama, the river descended sharply in a series of rapids, called Muscle Shoals (for the muscle power earlier required to canoe through them upstream). A canal bypassed the

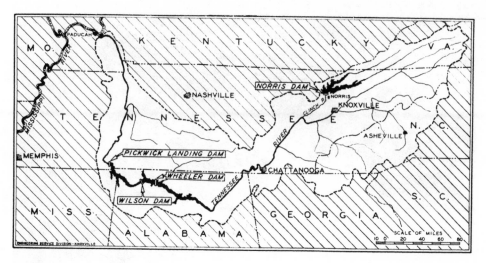

Figure 2.6
Tennessee River Valley (unshaded area) showing Wilson Dam and three other dams (Norris, Wheeler, and Pickwick Landing) built by the TVA in the 1930s. Courtesy of Tennessee Valley Authority.

Shoals in 1890 but alternating drought and floods made navigation of the river difficult and kept the Tennessee Valley isolated. Malaria was a menace along the riverbanks, and deforestation and soil erosion had caused agriculture in the region to deteriorate. The southern states were mostly rural and poor, and the Tennessee Valley and the hollows of the southern Appalachian Mountains were the South's most depressed area.[12]

A glimmer of change appeared in the early twentieth century. The sharp descent of the Tennessee River through Muscle Shoals made the site a potential source of water power to generate electricity. This power could also be used to benefit farming: organic fertilizers were no longer sufficient in quantity to meet the needs of American agriculture, and with new processes using electric power, inorganic nitrates could be manufactured to take their place. These nitrates could also be used to make explosives. During World War I (1914–1918), the U.S. Army relied on imported nitrates and worried that the supply might be cut. The Army Corps of Engineers hired the engineer Hugh Lincoln Cooper to build a dam at Muscle Shoals to supply hydroelectricity for nitrate production (figure 2.7). But construction of the dam, named for President Wilson, did not begin until 1920, and before it finished in 1924, President Harding offered to lease the structure to private electric utilities. Advocates of public control, led by U.S. Senator George Norris of Nebraska, prevented the lease, but Presidents Coolidge and Hoover vetoed bills introduced by Norris to allow the government itself to sell electricity from the dam.[13]

Figure 2.7
Wilson Dam, view from the downstream side. Courtesy of U.S. National Archives, Atlanta.

In his campaign for President in 1932, Franklin Roosevelt declared that the federal government had a right to produce and distribute electricity if power from private utilities was too limited or too costly.[14] In 1930, about one-quarter of the American people still lived on farms but only 10 percent of farmers had electric power, compared with the 84 percent living in urban and suburban areas who had power. The electric power industry had consolidated into a small number of holding companies, and critics argued that with more competition electricity could be less expensive and more widely accessible. Roosevelt pledged to bring change.[15]

The Tennessee Valley Authority

After taking office as President in March 1933, Roosevelt made development of the Tennessee River and its wider region a priority. On May 18, the President signed a bill introduced by Senator Norris that created the Tennessee Valley Authority, soon known by its initials, the TVA. The TVA Act charged the new agency to improve navigation on the Tennessee River, control flooding in its basin, and manage the dam and two nitrate plants that had been built at Muscle Shoals. The Act authorized the TVA to build new dams and sell to residents of the area any electric power not required for the agency's own tasks. Roosevelt hoped that electric rates charged by the TVA to residents would provide a "yardstick" to compare against the rates charged by private utilities, and that the comparison would encourage the latter to reduce their rates and serve more people. The Act created a three-member board of directors and gave the agency a budget of $50 million, supplemented by another $25 million in 1934.[16]

Soon after his inauguration, Roosevelt invited Arthur Morgan to serve as chairman and first director of the new TVA. Over the summer, Morgan asked Dr. Harcourt Morgan (1867–1950), president of the University of Tennessee (and no relation), to serve with him. Harcourt Morgan was an expert on local agriculture and its needs, and informally he represented the region in the TVA leadership. As the third director, Morgan appointed David Lilienthal (1899–1981), an attorney on the Wisconsin Public Service Commission, which regulated utilities. Wisconsin was the first state to institute statewide regulation of electric utilities, and Lilienthal had become known nationally for his vigilance on the Commission (figure 2.8).[17]

After some initial confusion, the new board divided their responsibilities. Harcourt Morgan took charge of TVA efforts to improve agriculture and land use, while David Lilienthal organized the distribution of electric power. Arthur Morgan directed everything else, principally the design and construction of new dams. Giving focus to Morgan's program was the idea of a multipurpose dam: a structure that would control flooding, facilitate river navigation, and generate electricity.[18] The Army Corps of Engineers had surveyed the Tennessee Valley in the 1920s for possible dam sites to generate hydroelectric power, and the Corps had chosen a site at Cove Creek on the Clinch River, an eastern tributary of the Tennessee. The Corps expected a private utility to lease the electricity once the dam was ready, but the project was unbuilt when President Roosevelt took office in 1933. The TVA resolved to complete the Cove Creek project and distribute the electricity itself. Perhaps to slight the Corps over its prior view of flood control, Morgan asked the Bureau of Reclamation to redraft the Cove Creek design, with a result that differed in only minor ways. Morgan adopted it, named the structure Norris Dam after the senator, and made plans to build four more dams in the 1930s.[19]

To build these dams, Morgan dispensed with the usual (and slower) practice of both the Bureau of Reclamation and the Corps of Engineers, in which federal engineers designed the structures and then contracted with private builders to do the construction. Instead, to employ people quickly and have closer control of the work, he trained and fielded his own organization of engineers, foremen, and laborers, as he had earlier for the Miami work. He recruited some of the engineers who had served on the Ohio project and added others from around the country. Some 39,000 men resident in the valley applied for jobs as laborers in the first year, of whom 9,173 were hired by June

Figure 2.8
The first TVA Board of Directors (*left to right*): Harcourt Morgan, Arthur Morgan, and David
Lilienthal. Courtesy of Antiochiana, Olive Kettering Library, Antioch College. Granville Hunt,
photographer.

1934 (figure 2.9). The number of employees rose to 16,457 by June 1935. At Morgan's
insistence, the TVA Act exempted the Authority from civil service laws, so as to permit
more flexible hiring of senior staff, and he was free to hire engineers and managers on
a basis of technical merit rather than political affiliation.[20]

The design of Norris Dam called for a straight (uncurved) concrete gravity dam
across the Clinch River. The dam reached 265 feet (80 m) in height and had a base
width of 208 feet (63 m). A concrete wall of 1,570 feet (478 m) in length and an added
earth embankment gave the dam a total crest length of 1,860 feet (566 m). A spillway in
the middle of the concrete wall allowed water to flow over the dam through gates that
could be raised or lowered as needed. The complex finished in July 1935, 21 months
after construction began (figure 2.10). Two more concrete gravity dams, Wheeler and

Figure 2.9
Residents listening to methods of employment with the TVA. Courtesy of U.S. National Archives.
Lewis Hine, photographer.

Pickwick Landing, finished in 1936 and 1938, respectively. Gravity dams at Guntersville in Alabama and at Hiwassee in western North Carolina finished in 1939 and 1940
with additional funding. The total cost of the five dams was $87,044,451.[21]

The new dams (and Wilson Dam) enabled the TVA to regulate the water level of the
Tennessee River, preventing sudden inundations from floods and allowing the release
of more water in periods of lower flow. Navigation as far as Knoxville became practical, and towns along the river could now trade with the outside world. By raising and
lowering the water level, in addition to spraying with insecticides, engineers could also
eradicate the breeding grounds of malarial mosquitos. The reservoirs submerged land
against the wishes of some displaced residents, and cemeteries and buildings had to be
relocated. But most of the valley's people welcomed the new civil engineering.[22]

The TVA dams also generated electricity. Some of the water behind each dam entered
underwater channels in the structure, called penstocks, that carried the water to powerhouses on the downstream side of the dam. Here the water spun *turbines*, shafts with

Figure 2.10
Norris Dam. Courtesy of U.S. Library of Congress.

blades on them. Atop each shaft, a magnetized disk or rotor spun inside a ring of coiled wire called a stator. The spinning generated an alternating current in the stator that could be fed to transformers. The transformers stepped up the voltage for transmission over long-distance wire lines, and transformers at the receiving end of these lines stepped down the voltage for local use (boxes 2.1, 2.2, and 2.3).[23]

The distribution of electric power began on a small scale at first, from Wilson Dam, which generated 184,000 kilowatts (1 kilowatt equaled 1,000 watts). About three-quarters of this power went to meet the needs of the TVA at the time. In disposing of the surplus, the TVA Act gave preference to nonprofit entities, such as municipal utilities and rural cooperatives. Low rates enabled residents receiving the power to buy and use electric irons, refrigerators, radios, water pumps, washing machines, and cooking ranges (figure 2.11). In 1940, all of the TVA dams had a generating capacity of 970,000

Figure 2.11
Electric cooking range supplied by TVA electric power. Courtesy of Franklin D. Roosevelt Library.

Figure 2.12
Land treated without and with new fertilizer. Courtesy of U.S. National Archives, Atlanta.

kilowatts. Thirty-eight percent went to public utilities and rural cooperatives, another 26 percent went to industrial firms, and almost 12 percent went to private utilities for resale. Ten percent went to other networks and about 7 percent was lost in transmission. The TVA's own needs by then absorbed only 5 percent of the power generated. Almost 400,000 residents of the valley, about one-third of the region's population, received TVA power in 1940, including 45,000 farmers.[24]

Harcourt Morgan did not make use of the two nitrate plants near Wilson Dam; by encouraging crop rotation in the valley, with leguminous vegetables replacing corn and other crops every other year, nitrogen in the soil could be replenished naturally. The greater need was to replenish phosphate in the soil, and the TVA built facilities to make a phosphate fertilizer with a higher nutritional content than standard fertilizers containing phosphate (figure 2.12). The TVA recruited farmers in the Tennessee Valley to turn their land into demonstration farms to test the new fertilizer and to test new crops and other new farming methods. The Authority also promoted the conversion of some land to grass for grazing to conserve the topsoil, and the agency introduced new methods to achieve a more sustainable rate of timber cutting.

Box 2.1

Water Power from Wilson Dam I

The U.S. Army Corps of Engineers built Wilson Dam across the Tennessee River near Muscle
Shoals, Alabama, in 1924. The dam came under the Tennessee Valley Authority in 1933 and
stands 137 feet (41.7 m) high and has a length of 4,541 feet (1,384 m) from shore to shore.
Bladed shafts called turbines used the water to generate electricity.

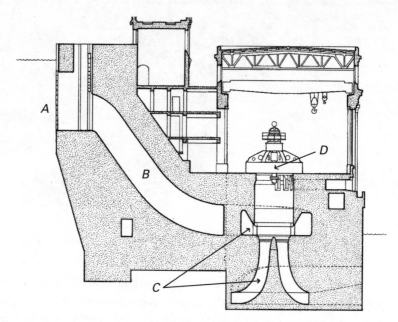

Cross-section of penstock and turbine at Wilson Dam.

To produce electricity, water behind the dam, *A*, entered a descending channel called a
penstock, *B*, and turned the blades, *C*, attached to the turbine shaft. In the housing, *D*, the
shaft spun a magnetized disk, the rotor, inside a ring of coiled wire, the stator.

Spinning the rotor inside the stator generated electricity in the stator that could be sent
out by wire. In this way, the turbine converted the mechanical power of the moving water
into electric power. Water exited the turbine from the bottom.

Source: Image courtesy of the U.S. Library of Congress. Image No. HAER ALA,17-MUSHO,2A- (sheet 1 of 1).
Original drawing made by Chris Payne. Letters *A*, *B*, *C*, and *D* added.

Box 2.2
Water Power from Wilson Dam II

The mechanical power of water entering a turbine (P_{in}) is the product of three quantities: the *flow* of the river (Q) in cubic feet per second, the *density* of the water (w) in pounds per cubic foot, and the *head* of the water in feet (H), or the height difference between the water surface behind the dam and the water level leaving the turbine.

Wilson Dam originally had eight turbines installed. In four of them, water entering each turbine had an average flow of 3,875 cubic feet per second. The density of water is 62.4 pounds per cubic foot.[a] The head at Wilson Dam is 92 feet but power equivalent to 1 foot was lost to friction, giving a net head of 91 feet. Dividing by 550 gives the horsepower (hp):

$$P_{in} = Q\,w\,H\,/\,550$$

$$P_{in} = (3{,}875)\,(62.4)\,(91)\,/\,550 = 40{,}000 \text{ hp}$$

The power of the water going out of a turbine (P_{out}) is the product of the power going in (P_{in}) multiplied by a rating, the percentage not lost from friction in the turbine itself.[b] The above four turbines at Wilson Dam had a rating of about 87 percent, giving the power out of each turbine as 35,000 hp:

$$P_{out} = (P_{in})\,(0.87)$$

$$P_{out} = (40{,}000)\,(0.87) = 35{,}000 \text{ hp}$$

The other four turbines had power outputs of 30,000 hp each. Converting the horsepower of the turbines into electricity made the turbines electric generators (see box 2.3).

Sources: Data from the TVA. Flow rate is an estimate and numbers for horsepower have been rounded slightly. *Notes:* (a) Using the specific weight of water at the location (62.22 pound-force/foot³) instead of density would return a slightly smaller number for horsepower. (b) The rating of these turbines was 87.75 at 92 feet gross head and full load. The four turbines detailed above are Units 5–8. Units 1–4 had the outputs of 30,000 hp.

These new land use practices began to reverse soil erosion, restore forest and farm-land, and improve the productivity of the land. Harcourt Morgan worked through existing federal, state, and local agencies and through local colleges to promote these changes, rather than dictate change from above or compete with other public authori-ties. Demonstration farms soon spread to twenty-two states outside the Tennessee Val-ley, giving the TVA a national influence.[25]

Morgan versus Lilienthal: Conflicting Visions

The TVA dams were gravity designs, and in the generation and distribution of hydro-electric power the agency followed standard engineering practices. The Authority's work in farming and forestry was innovative but gradual in its effect. What was radical

Box 2.3
Hydroelectricity from Wilson Dam

Transformers in switchyard *(left)*; transmission lines from switchyard *(right)*.

Water going through four of the eight turbines at Wilson Dam produced about 35,000 hp each. Each turbine generator converted 1 unit of horsepower into 0.746 kilowatts of electric power (1 kilowatt equals 1,000 watts), and each generator could convert its horsepower into kilowatts with an efficiency of 97 percent.

Multiplying 35,000 × 0.746 and the result by 0.97 gave each generator an output of about 25,000 kilowatts. The other four generators produced 21,000 kilowatts each, giving the dam a total electric power output of 184,000 kilowatts.

Before sending out this power over transmission lines, the switchyard at Wilson Dam had to change the relation of voltage and current in the electricity (the flow of electricity is the current and voltage is the pressure of this flow).

The electric power out of a generator (P) is equal to the voltage (V) × the current (I), or $P = VI$. Power lost in transmission increases with the size of the current, so by stepping down the current and raising the voltage, very little power is lost when sent over long distances. Transformers at the generating station were devices that could step up the voltage and step down the current for transmission. Transformers at the receiving end could step the voltage back down and the current back up for local use.

Generating capacity has been added to Wilson Dam since the 1930s, giving the dam an electric power output today of about 663,000 kilowatts.

Sources: Images courtesy of the U.S. Library of Congress. Alfred T. Palmer *(left)* and Arthur Rothstein *(right)*, photographers. Data from the TVA. The other four generators operated with an efficiency of 95.5 percent.

about the TVA was its mission to change an entire region. Other regional agencies in the United States addressed more specific problems: the Port of New York Authority tried to relieve traffic congestion in the New York metropolitan area; the Bonneville Power Administration supplied electric power to the Pacific Northwest; and the Metropolitan Water District brought water to southern California.[26] The Tennessee Valley Authority had a more comprehensive aim: to control flooding and river navigation, to improve how farmers worked the land, and to produce electric power, all to give the region a better standard of living. A dedicated engineering organization and workforce restructured the landscape and began to change how its people lived and worked.

The three TVA directors disagreed, however, over longer-range goals. Arthur Morgan's aim was to set an example of small towns close to nature that could maintain a way of life superior to the life of huge impersonal cities. Alongside Norris Dam, he built houses for construction workers, in a setting called Norris Village, as a model community that he hoped would continue after the dam had been built (figure 2.13). Morgan encouraged craft industry to develop but he also wanted the residents to benefit from modern engineering, and he equipped the houses, a school, and a community center to use electricity. Morgan did not see the renewable character of hydroelectricity as a principal advantage, but his vision resembled what might today be called a more sustainable society. His overriding concern, though, was ethical. He imposed a code of conduct on TVA personnel that enjoined each employee to be honest and truthful, suppress selfish ambition, live modestly, and cooperate for the good of all. He wanted the Tennessee Valley to set an example to the rest of America of people living together in a spirit of cooperation that he hoped others would emulate voluntarily. Arthur Morgan wanted the "yardstick" of TVA electric power to show how people could be served at a lower cost, but he opposed competing with private utilities for the same customers. He hoped that the utilities would reduce their rates when they saw that lower rates in the area demonstrated by the TVA could raise demand.[27]

David Lilienthal believed that the private utilities in the region would never reduce their rates unless faced with competition. He built transmission lines to offer customers throughout the valley a public supply of electric power at lower rates, in order to force the private utilities to lower their prices or go out of business. As demand for electricity grew at the lower rates, Lilienthal believed, it would also increase demand for electrical goods and help revive the wider national economy. The TVA was able to market electricity to rural users at about half the price charged by private utilities to urban customers.[28] In 1933, Arthur Morgan reached a three-year agreement with Wendell Willkie, who led Commonwealth and Southern, the dominant holding company for private

Figure 2.13

Norris Village houses. Courtesy of U.S. National Archives, Atlanta.

electric utilities in the South. The TVA agreed to limit its marketing to an area supplied by Wilson Dam if the agency could obtain some privately held distribution facilities to demonstrate its "yardstick" approach in this area. Lilienthal acceded to the agreement but continued to build electric power facilities attached to the other dams.

Harcourt Morgan's goal was to improve the agrarian life of the region in certain ways, with the cooperation of the local society. However, he backed Lilienthal against the other Morgan as the private electric utilities in the region challenged the very existence of the TVA. In 1934, shareholders of the Alabama Power Company, a firm held by Commonwealth and Southern in the area served by Wilson Dam, filed a lawsuit to enjoin the TVA from acquiring the facilities of the smaller company. The shareholders argued that the federal government had no right to use taxpayer funds to create a public competitor to private industry. Lilienthal waged a vigorous legal defense, and the case went up to the U.S. Supreme Court, which ruled in 1936 that selling surplus electric power from Wilson Dam, power the TVA did not need for itself, was a way of selling surplus federal property. The court ruled that the federal government had always had the right to sell such property.[29] In response, Willkie challenged in court the sale of lower cost electric power from all of the other TVA dams. The case again went up to the Supreme Court, which again found in favor of the Tennessee Valley Authority in early 1939. As a result, the private utilities in the valley conceded defeat and sold most of their assets to the TVA. Willkie went on to win the Republican nomination for President of the United States in 1940, only to lose to Roosevelt in the fall election of that year.[30]

Disagreement over electric power and other matters divided the three-member TVA board and eventually made their working together impossible. Lilienthal's appointment expired in 1936, and Arthur Morgan tried to persuade Roosevelt not to reappoint him. However, the President did not want an embarrassing rupture during an election year in an agency that symbolized his presidency, and he reappointed the co-director. Relations on the board deteriorated until finally, in 1938, Arthur Morgan made charges against the other two board members in public that he could not substantiate, and Roosevelt removed him. The visions of both Morgan and Lilienthal had appealed to President Roosevelt, but in the end, he had to choose between them. Harcourt Morgan served as chairman until 1941, when David Lilienthal became the sole director.[31]

By the end of the 1930s, Arthur Morgan's social vision for the TVA had also failed. The residents of Norris Village could not subsist on craft industry when work on Norris Dam finished in 1936, and commuters from Knoxville eventually took their place. Harcourt Morgan's program fell short in that it did not benefit the very poor. The farmers most aided by the TVA were those who owned their land, not tenants and sharecroppers, and the beneficiaries of the agency included few of the people living in the more remote mountain hollows of southern Appalachia. The TVA had pledged not to engage in racial discrimination and gave jobs to 2,105 African Americans by June 1935, paying wages equal to those of white workers for the same work. But African Americans lived in segregated housing and were kept in less skilled occupations.[32]

Despite its shortcomings, the TVA of the 1930s was a radical public innovation. Arthur Morgan's engineering organization proved that immense projects could be built efficiently and quickly to create a modern network of water and power over a multistate region. The TVA's degree of independence from federal and state control gave it a unique sense of mission, attracted dedicated and highly capable engineers, and gave employment to thousands of people. The deterioration of forest and farmland began to end, and David Lilienthal's power program brought electricity to rural areas and small towns over a wide area, becoming the model for the nationwide Rural Electrification Administration set up in 1935. By 1946, 670,000 people in the Tennessee Valley received electricity from the TVA, and most of the region received power in the years that followed.[33]

After leading the TVA during America's participation in World War II from 1941 to 1945, David Lilienthal accepted appointment in 1946 as first head of the Atomic Energy Commission (AEC), which took over the Manhattan Project, the wartime military program to develop the atomic bomb (see chapter 4). As the Cold War began, though, the AEC had to devote nearly all of its early work to producing nuclear weapons, not to developing nuclear energy for peacetime use, as Lilienthal had hoped it would be able to do. He returned to private life in 1950.[34]

The Postwar TVA

As the war drew to an end, the TVA inspired some in Congress to propose regional authorities in other parts of the country where multipurpose dams had been built and where additional ones were planned. Lilienthal had argued during the war that the TVA was an advance in American democracy. A regional authority such as the TVA was accountable to the federal government in Washington, he argued, yet was independent of detailed management from afar and could be closer to the people that it served.[35] Harcourt Morgan had muted the possible tension between the TVA and local institutions of government by working through the latter as much as possible. But Lilienthal's emphasis on the TVA's independence, and his prewar clash with private industry, underlined the potential threat of the TVA model to private interests elsewhere in the country and also to more centrally run federal agencies that had responsibilities of their own over rivers. In the late 1940s, opponents in the federal government and in the affected states blocked the creation both of a Columbia Valley Authority in the Pacific Northwest and a Missouri Valley Authority in the upper Plains states, and enthusiasm for regional agencies of this kind subsided in the early 1950s.[36]

The TVA itself faced an uncertain future after the war. Following Lilienthal, a single director replaced the three-member board, and service as an electric utility became the primary mission of the agency. However, the hydroelectric potential of the region was nearly used up by 1945, and existing dams could not meet the rising postwar need for electricity in the Tennessee Valley. Half of this demand came from the federal government, chiefly to power the making of nuclear weapons at Oak Ridge, Tennessee. To meet the new demand, the TVA built power plants that would burn coal. Conflict revived with private utilities that feared additions to the agency's generating capacity, and after 1952, President Dwight Eisenhower opposed federal funding for TVA expansion. Finally, in 1959, Congress authorized the agency to raise its own money privately through the sale of bonds, which enabled the TVA to complete a series of coal-burning power plants in the 1960s and early 1970s, along with several nuclear power plants.[37]

The TVA named (with some unintended irony) its most ambitious fossil-fueled power plant after the nearby village of Paradise, Kentucky (figure 2.14). In each of two huge generating units, coal delivered to the Paradise power plant burned in a huge boiler to produce steam that spun a turbine (box 2.4). Each unit consumed 306 tons of coal per hour and could produce 650,000 kilowatts of electric power. When finished in 1963, the plant was the largest coal-fueled power station in the world. A third generating unit added in 1970 enabled the plant to serve almost a million homes and raised its annual consumption of coal to about 7 million tons, or about 1 percent of the energy used nationally to generate electric power at the time.[38]

However, even the best coal plants were able to convert less than half of the energy in coal into usable electric power, and the shift to coal adversely affected the natural environment. The chimneys at Paradise and other steam plants vented fly-ash and gases from burning coal, and while scrubbers and other devices reduced these emissions, they could not eliminate them. To supply Paradise and other coal-fueled plants, TVA contracted with private mining companies that made use of open-air excavation. This kind of mining badly damaged the landscape, undoing much of the agency's earlier work to bring soil erosion under control. A new dam at Tellico, on a tributary of the Tennessee River, threatened a small endangered fish, the snail-darter, causing a

Figure 2.14
Paradise power plant. Courtesy of Tennessee Valley Authority.

national controversy in the 1970s.[39] The economic advantages enjoyed by the early TVA also diminished. Its nuclear plants proved unexpectedly expensive to build, and the price of coal began to rise in the late 1960s, requiring the agency to raise the rates it charged for electricity after three decades of charging low prices.[40]

In 2013, the TVA pledged to reduce the percentage of its electricity generated by coal from 38 to 20 percent by 2020. In 2017, the agency shut down two of the three coal-fueled generating units at the original Paradise plant and in their place began generating power from a new Paradise plant nearby that burns natural gas. The gas enters each of three horizontal gas turbines, where it mixes with compressed air and ignites. The combustion in each turbine spins a shaft, with a rotor and stator at the end, to produce electricity. The plant recycles the heat by sending it to a steam generator attached to each turbine, where the heat turns water into steam. The steam then turns a fourth turbine that works with steam, producing additional electricity, and a condenser turns the steam back into water. The original Paradise plant turned coal into useful electrical energy with an efficiency of 40 percent. The new Paradise plant burns natural gas with an efficiency of 60 percent and emits only half as much carbon dioxide as the coal

Box 2.4

The Original Paradise Power Plant

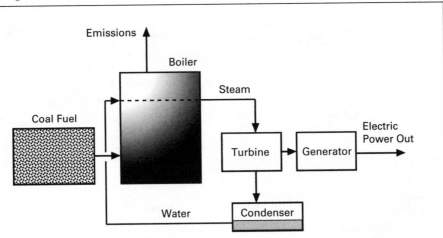

Diagram of a generating unit at the Paradise power plant.

Finished in 1963, the TVA power plant near the village of Paradise, Kentucky, was the largest fossil fuel plant in the world at the time. Each of its two massive generating units received a coal supply fed into a boiler. A pipe loop sent water through the boiler, turning it into steam that spun a turbine, generating electricity. The steam left the turbine and entered a condenser, where it turned back into water to repeat the cycle.[a]

One unit of horsepower has a force of 550 foot-pounds per second, or 33,000 foot-pounds per minute, and burning 1 pound of coal produces a force of 8.1 million foot-pounds.[b] Each generating unit consumed 612,000 pounds of coal every 60 minutes. Dividing the force in 1 pound of coal by 33,000, and multiplying by the 612,000 pounds of coal divided by 60, gives the potential horsepower in the coal:

$$\frac{8,100,000}{33,000} \times \frac{612,000}{60} = 2,500,000 \, \text{Hp}$$

Each generating unit produced 650,000 kilowatts of electric power, equivalent to 871,000 hp.[c] Dividing this number by the 2,500,000 hp in the coal gave the efficiency with which the generating unit converted the energy in the coal into useful energy: 34 percent. The efficiency of coal plants has risen since then only to about 40 percent.

Source: Reed A. Elliott et al., "TVA's Paradise Steam Plant," *ASCE Journal of the Power Division* (May 1962): 89–119. *Notes:* (a) Cooling towers on site (not shown in diagram) also helped cool the water. (b) Coal has a heat content of about 10,400 British thermal units (BTUs) per pound. One BTU equals 778 foot-pounds of force. (c) One kilowatt equals 1.34 hp.

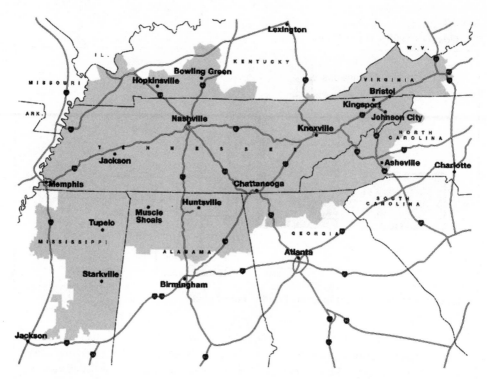

Figure 2.15
TVA service area today (shaded). Courtesy of Tennessee Valley Authority.

plant. Recognizing a need to go further, the TVA has also begun to increase its use of energy from the sun, wind, and other renewable sources (besides its original dams) to supply electricity in its service area (figure 2.15).[41]

The Founders in Retrospect

The small communities that Arthur Morgan idealized could not have built the hydro-electric dams that Morgan created, and the renewable energy became part of a larger and more conventional utility. But in his engineering organization Morgan challenged the America of his time to manage a public service and the larger society with a stronger ethical sense, and his notion of small community challenged the impersonalism in much of modern life. David Lilienthal's vision of public power evoked opposition that limited the TVA model to its original region, but he proved that a public service could meet a pressing social need, for electric power, when private industry failed to do so.

The TVA also succeeded because the public investment in its engineering was economical, quickly gaining more users of its electricity with a lower price. To the extent that it can move toward nonpolluting and more sustainable sources of energy, and still deliver affordable power, the agency is returning to its roots.

Arthur Morgan's first major accomplishment as an engineer, flood control in the Miami Valley of Ohio, has stood the test of time. During the 1920s, the Army Corps of Engineers still maintained levees on each bank of the Mississippi River to control flooding. A buildup of silt gradually raised the floor of the great river, though, and in 1927, rainfall followed by a great flood broke the levees in several places, killing hundreds and making three-quarters of a million people homeless. The Corps finally added diversion channels along the lower length of the river.[42]

Two threats to settlement along the Mississippi remained: flooding along the tributaries of the river in the upper Midwest, where cities and towns nestled close to riverbanks, and surges of seawater at the mouth of the Mississippi on the Gulf of Mexico during tropical storms. After heavy rains in 1993, the tributaries and the river itself flooded in the upper Midwest, and in 2005 Hurricane Katrina overwhelmed the city of New Orleans. Land use better suited to periodic flooding has been difficult to advance in areas that are now densely settled.[43]

Another great investment in America's twentieth-century public infrastructure, alongside dams to control rivers and produce electric power, was the building of paved highways. A national highway network, substantially completed by the 1980s, enabled a majority of Americans to live in suburbs and embrace a way of life that depended on private cars. Instead of civil engineering projects that were local or regional in extent, and completed in a matter of years, the highway network was a program that stretched over many decades and covered the entire United States.

3 Highways and Skyscrapers: MacDonald and Khan

The most popular exhibit at the New York World's Fair of 1939–1940 was the General Motors pavilion. Its theme of "Highways and Horizons" featured an attraction, the Futurama Ride, that looked forward to the American landscape in the year 1960. The ride circled a huge diorama that depicted a country of cities with tall skyscrapers interconnected by superhighways.[1] This vision of the future came true by 1960 as could be seen from the windows of an airplane landing near any major American city.

The greatest American infrastructure project of the twentieth century was a national highway network. The first motor vehicles appeared in America in the 1890s and the mass production of automobiles began in 1913, employing hundreds of thousands of workers and making cars affordable to millions. Just as important, the discovery of abundant petroleum reserves in the United States, and the engineering of new processes to refine gasoline, enabled the oil industry to supply these vehicles with fuel. Automobile registrations grew as a result from just four in 1895 to eight million in 1920.[2]

During the 1920s, the driving public began to demand better roads, and the federal government also saw a military defense need for them. The Bureau of Public Roads under Thomas MacDonald planned and helped the states construct a network of what came to be known as United States Highways. As the number of motor vehicles continued to grow and as cars reached higher speeds, these roads became inadequate. During and after World War II, MacDonald's Bureau made plans to build a network of express highways, which would be roads with no stoplights on them. Construction began in 1956 to build this network, the Interstate Highway System.

The innovation of structural steel in the late nineteenth century made it possible for urban buildings to rise much higher. These buildings allowed more people to live and work in cities, with the tallest "skyscrapers" rising first in Chicago and then in New York. The 1960s brought a second revolution in building, as skyscrapers rose to 100 floors or more, at less cost for buildings this tall, as a result of a radical innovation by the engineer Fazlur Khan.

By the early twenty-first century, cities reached into the sky and motor vehicles and new highways interconnected the country. In recent years, concerns over the natural environment have raised questions about the future of an automotive civilization. While both have helped to define this civilization, machines and structures reflect different values that a more sustainable future may need to have in better balance.

Modern Road Building

Although some streets were made of cobblestone, nearly all roads in the early nineteenth-century United States were unpaved. Travel on them was by horse, horse-drawn carriage, or on foot. In the 1820s, civil engineers began to harden roads with a method invented by John McAdam in Britain, who found that crushed stone would solidify on an earth foundation under the weight of repeated use. The surface could drain rainwater as long as the middle of the road was slightly higher than the roadway's edges, allowing the water to flow into ditches alongside. However, rivers and canals and then railroads came to carry most of America's passengers and freight, and nearly all American roads were still unpaved by the end of the century.[3] Bicyclists launched a "Good Roads" movement in the 1890s but made little headway. Henry Ford designed his 1908 Model T, the vehicle that began assembly-line mass production of cars, to drive on the earth roads that were typical of the United States at the time.[4]

The Model T proved a turning point, though, by creating a much larger driving public. Mud caused this public to demand better roads (figure 3.1). After 1900, American states began to spend money to improve roads, usually by covering earth roads with loose gravel. The Federal Aid Road Act of 1916 provided federal matching funds to each state to improve its roads and build more paved ones. The Act allowed states to choose which roads to build or improve but required the states to maintain them. To qualify for matching funds, each state had to have a highway department under a professional civil engineer, and the eighteen states that did not already qualify soon did so.[5]

To build a paved road, a public authority first acquired the strip of land that the roadway would follow, with some compensation to owners for right-of-way that went through private property. Civil engineers and road workers then cleared the ground and made any uphill or downhill grades more gradual. In rough or mountainous terrain, clearing often required demolition of rock. Laying a foundation was next, consisting usually of one or more layers of sand and crushed stone on compacted earth. Finally, engineers and workers paved the road with an acceptable width and a slight rise

Figure 3.1
An unpaved Iowa road in 1919. Courtesy of and © Iowa Department of Transportation. All rights reserved.

in the middle to facilitate drainage into ditches along each side. The outer sides of the road also had to have shoulders so that vehicles could pull over to stop.

A paving material better able to carry motor vehicles, concrete, began to enter use in the early twentieth century. In the 1820s, Joseph Aspdin and his son William, bricklayers in Britain, invented a key ingredient called Portland cement (box 3.1). This cement consists of crushed limestone, a common rock, and clay, a common soil. Subjecting both to high heat forms a mixture that cools into pellets. These are mixed with a smaller amount of another mineral, gypsum, and crushed to form a powder. To make concrete, workers mixed the powder with water to create a cement paste, which formed a binding agent when combined with an aggregate of sand and crushed rock. The result, after pouring into wooden molds to harden, was concrete (figures 3.2 and 3.3).[6]

A different way to make concrete was to use asphalt instead of Portland cement as the binding agent. Asphalt was a residue left after lighter chemicals, such as gasoline,

Box 3.1

Making Concrete

Concrete is an artificial stone made by mixing a binding agent with water and then adding an aggregate of sand and crushed rock. There are two binding agents used in modern paved roads: Portland cement or asphalt.

Portland cement begins with a mineral, calcium carbonate ($CaCO_3$), found in limestone, and silicon dioxide (SiO_2), found in a kind of soil called clay. Cement makers crush the limestone and mix it with the clay (in a ratio of about 3:1), and then subject the mixture to high heat, about 2700°F (1482°C).

In a typical reaction, three molecules of limestone and one molecule of clay react to form one molecule of tricalcium silicate, which engineers call alite, and three molecules of carbon dioxide:

$$3CaCO_3 \quad + \quad SiO_2 \quad \rightarrow \quad Ca_3SiO_5 \quad + \quad 3CO_2$$

| calcium carbonate (limestone) | silicon dioxide (clay) | tricalcium silicate (alite) | carbon dioxide (emitted to atmosphere) |

Smaller amounts of other minerals are added to the alite and crushed to form a cement powder. Workers then mix the cement powder with water, which creates calcium silicate hydrate, or cement paste, and add calcium hydroxide or builder's lime:

$$2Ca_3SiO_5 \quad + \quad 6H_2O \quad \rightarrow \quad Ca_3Si_2\,O_{10}H_6 \quad + \quad 3CaO_2H_2$$

| alite | water | cement paste | builder's lime |

Mixing the paste and lime with an aggregate of crushed rock and sand forms concrete in its fluid state. The fluid goes into a wooden formwork to set or harden.

A different way to surface a road is to use asphalt instead of cement paste as the binding agent. Asphalt is a residue left after other chemicals, such as gasoline, have been refined from crude petroleum. Asphalt is mixed with crushed limestone and poured onto the prepared underbed for a road. Rollers then compact the asphalt mixture to form a smooth roadway.

Seasonal changes in temperature can cause paving material to crack and loosen, and paved roads must be repaired every year. The manufacturing of cement and the refining of oil are also sources of carbon dioxide emissions into the atmosphere. Broken concrete can be recycled to reduce the need for new concrete.

Source: Concrete-making information from David P. Billington, Sr.

Figure 3.2
Placing concrete for a paved road in the early 1920s. Courtesy of and © Iowa Department of Transportation. All rights reserved.

had been refined from crude petroleum. Asphalt roads were black in color, while roads made of Portland cement were beige in color. Both kinds of surface were concrete, but the term "concrete" customarily refers to the kind made with Portland cement, while "asphalt" refers to the other. Road building required steam-powered shovels and rollers, which were invented in the nineteenth century. The ability to mix asphalt and aggregate on site, and the coming of movable cement mixers for Portland cement, made it easier in the 1920s and 1930s to lay pavement quickly and economically.[7]

Roads and Research

During America's participation in World War I (1917–1918), the federal government decided to relieve railway congestion by using heavy motor trucks to move war supplies from the Great Lakes region to the East Coast. This proved disastrous when existing paved roads cracked and broke up under the weight and volume of the truck traffic.[8]

10,529

Figure 3.3
Finished paved road in 1923. Courtesy of and © Iowa Department of Transportation. All rights reserved.

To rebuild the nation's roads, President Woodrow Wilson turned to a civil engineer from Iowa, Thomas Harris MacDonald (1881–1957) (figure 3.4). Born in a log cabin, MacDonald was a 1904 graduate of the Iowa State College (now Iowa State University) at Ames, and he became the first state highway engineer of Iowa in 1907, riding horseback to promote and help make better roads. In 1918, he was elected president of the American Association of State Highway Officials (AASHO), the national group of state highway engineers, and the following year the President appointed him to head the federal Bureau of Public Roads, which oversaw the nation's roads. MacDonald would direct the Bureau until 1953 and would lay out the nation's federal highway network. Although a man of unusual formality, whom subordinates called "Chief," he recognized a need for dramatic action in response to the wartime crisis.[9]

MacDonald blamed the wartime failure of the roads on a lack of scientific knowledge. Engineers needed, he believed, to have a more precise understanding of how

Figure 3.4
Thomas H. MacDonald. Courtesy of U.S. Federal Highway Administration.

roads and vehicles interacted.[10] In this new approach, MacDonald followed what he and his colleagues believed to be the scientific trend in all of engineering at the time. After rising to greatness, the telephone, steel, petroleum, and electrical industries had established research laboratories and hired scientists as well as engineers to staff them. The work of these labs consisted mainly of improvements to technologies, most of which had come into being without science as a direct stimulus. But improvements were often complex and required experts with specialized training in mathematics and science as well as engineering. The new industrial laboratories made the electric light bulb, alternating-current electric power, and long-distance telephony more efficient.[11]

MacDonald began a research program that brought success at first. Studies showed that paving roads with concrete made of Portland cement or asphalt was better than trying to repair older roads made of crushed stone. Driving trucks across sensors embedded in pavement proved that more damage resulted from unevenness in the paving than from the weight of the vehicle. Most importantly, tests found that tires filled with air were less damaging to roads than the solid tires then in use, and motor vehicles began to shift to inflatable tires as a result. On paved roads with two lanes, one in each direction, the Bureau of Public Roads showed a need to increase the thickness of pavements from four inches to six inches or more, and to make such roadways at least 20 feet wide.[12]

However, in the manner of science, the Bureau wanted to isolate and study each component of certain problems. To measure the impact force and the reaction of a pavement slab, the Bureau began to use separate testing machines for each measurement. This approach gave more exact data but removed the testing from actual road conditions. Test results such as these won praise for their laboratory rigor but were of less practical value to engineers. The Bureau also tried to study soil foundations to discern general principles for engineering. On actual roads, though, soil foundations varied too much for the Bureau to provide the kind of guidance it sought. The Bureau's work advanced road building in ways that would soon transform the country, but by the end of the 1930s it became clear that road engineering would never be an exact science.[13]

State highway engineers in Illinois made important findings in a different way in the early 1920s by conducting trials on a roadway loop known as the Bates Test Road. Driving heavy trucks around the loop over many different kinds of surfaces for several months confirmed that concrete paving was the best surface and also found that drivers tended to drive closer to the outer edge of a two-lane road, so as to be clear of traffic in the other direction. Roads needed to be stronger at their outer edges as a result.[14]

A National Highway Network

Under the direction of Thomas MacDonald, the Bureau of Public Roads built a reputation as an engineering organization imbued with high professionalism and an ethic of public service, similar to the early TVA under Arthur Morgan. Bureau engineers assisted and inspected federally funded roadwork around the country and helped bring uniform standards to road building in the United States, in partnership with state highway departments, which accepted Bureau advice in the interest of technical efficiency and safety.

In 1921, MacDonald asked the U.S. Army to designate roads that the United States would need to be paved in the future for national defense. General John J. Pershing,

Figure 3.5
Map of United States Highways in 1926, showing greater density in the more populated areas.
Courtesy of U.S. Federal Highway Administration.

commander of U.S. forces in Europe during World War I and now army chief of staff, proposed a road network to speed travel in the more populated and industrialized parts of the country. The roads would also enable the army to move forces from one coast to the other quickly. MacDonald secured the Federal Highway Act of 1921 to give states new matching funds to build or improve roads, so that they could meet federal standards and connect across different states. Although each state was free to choose the roads to be funded as part of the national highway program, most followed the proposed layout (figure 3.5).[15]

The national network, called United States Highways, consisted of paved two-lane roadways, one lane for each direction of traffic. Intersections were allowed, along with access from driveways, which limited highway vehicle speeds to 25–35 miles per hour (40–56 km/hr). MacDonald chaired a committee of federal and state highway engineers

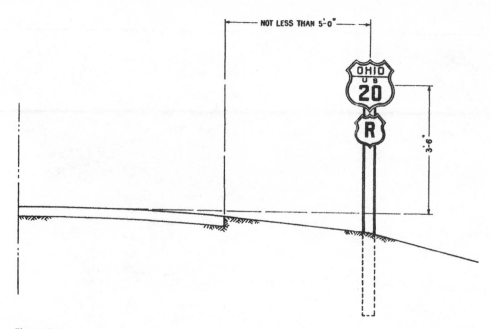

NOT LESS THAN 5'-0"

OHIO
U S
20

R

3'-6"

Figure 3.6
A U.S. Highway road sign standard in 1927. Courtesy of American Association of State Highway and Transportation Officials.

that gave these roads route numbers: those going north to south received odd numbers and those going east to west had even ones. Odd-numbered routes began along the East Coast, starting with U.S. Route 1, and even-numbered ones began close to the Canadian border with U.S. Route 2. New signs identified these routes (figure 3.6).

From 1921 to 1928, the mileage of paved concrete roads made with Portland cement or asphalt grew from 22,190 to 67,399 miles (35,711 to 108,468 km), and the mileage of all surfaced roads (mostly surfaced with gravel) doubled from 387,760 to 626,137 miles (624,039 to 1 million km), leaving 2,390,000 miles (3.8 million km) of earth roads. The states paid for their share of road costs through driver license fees and taxes on the retail sale of gasoline.[16]

New road-building in the 1920s helped suburbs grow. Americans had begun to move into smaller communities outside the larger cities in the 1880s, when electric streetcars and commuter railroads made it possible for the wealthier to work and shop in city centers and live in suburbs. Streetcars improved urban life as well, by eliminating the wastes associated with horses and by allowing people on foot to travel beyond their immediate neighborhoods. In the 1920s, installment plans made houses and cars easier to finance with borrowed money, and with cars more people were able to live in suburban communities farther away from streetcar and railway stops. Private streetcar companies began to decline in the 1910s, mainly because their costs rose more quickly than they could raise their fares, and after 1920 motor vehicles gave them competition. In the 1930s, more flexible bus services began to replace street railways.[17]

Modern Express Highways

During the Depression of the 1930s, America strengthened its public works infrastructure. In addition to water and electric power projects, the federal government funded road repairs and the construction of new highways. Greater federal spending on roads began under President Hoover as a way to give useful work to the unemployed, and after 1933 President Roosevelt and Congress gave roadwork half of a much larger budget for relief employment. From 1930 to 1940, the miles of surfaced roads in the United States rose from 694,000 to 1,367,000 (1.1 to 2.1 million km). The share of surfaced roads made with Portland cement or asphalt went from 84,000 to 153,000 miles (135,184 to 246,229 km).[18]

Motor vehicles brought two new challenges in the 1930s. First, vehicle traffic moving in and out of cities continued to increase, along with street traffic in the cities themselves. Traffic congestion threatened to overwhelm the nation's urban areas if the number of automobiles continued to grow. Under the federal highway acts of 1916 and 1921, federal aid went to road-building outside cities, which were responsible for their own roads and streets. In 1933–1934, Congress made urban areas eligible for federal highway funding, although only a few new urban highways were built before World War II.[19] The second challenge emerged outside of cities. Automobile engines grew more powerful in the 1930s, and with advances in the chemical engineering of gasoline, cars could reach speeds of 60 miles per hour (96 km/hr) and higher. The nation's earlier investment in long-distance highways, designed for speeds of 25–35 miles per hour, could not safely accommodate vehicles at these higher speeds. Americans regarded higher speed as a necessity, and public engineers believed that roads capable of carrying vehicles at higher speeds would also be essential to national defense.[20]

Highway engineers recognized that driving at higher speeds would be safer on "express highways" that motor vehicles could enter and leave only from access roads or ramps. Intersecting cross-traffic would go over or under these highways, and on those with four lanes a wall or grass median strip would separate opposing lanes (figure 3.7). By the 1930s, engineers had a formula to calculate roadway stopping distances that

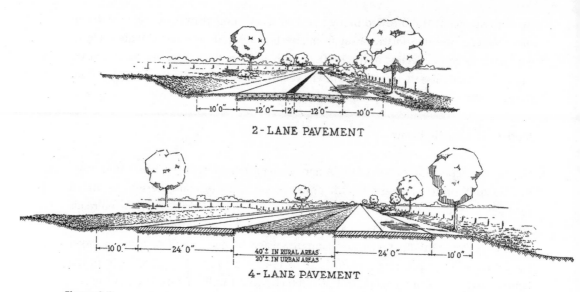

Figure 3.7
Express highway profiles proposed in 1939. Courtesy of U.S. Federal Highway Administration.

made expressway design safer (box 3.2). Highway engineers also knew that entering a curved turn or ramp at high speed would produce a centrifugal force that risked throwing the car off the road. To reduce this danger, engineers imposed a slower speed limit on a curved turn and designed the turn surface to have a rise, called a super-elevation, that made the outer side of the curve higher than the inner side.[21]

However, MacDonald and the Bureau of Public Roads did not see a need for expressways in rural areas. Believing that traffic counts in the present were the way to predict traffic in the future, the Bureau found that nearly all present-day traffic was in or close to cities. MacDonald and his engineers urged more urban expressways. As a result, the Bureau incurred its first serious political criticism by states that wanted better roads in rural areas, and advocates of toll roads added a criticism of their own.[22]

Toll roads were private highways that charged for access, such as the Merritt Parkway in Connecticut, which opened in 1938 to provide a scenic alternative to U.S. Route 1, which ran parallel to the parkway but south of it. The parkway's success disproved Thomas MacDonald's belief that drivers would not pay to travel on roads if comparable roads nearby and going in the same direction were free.[23] At the urging of President Roosevelt, Congress asked the Bureau of Public Roads to study the possibility of building a national network of express highways and to recommend whether they could be financed with tolls. The Bureau wrote a 1939 report, *Toll Roads and Free*

Box 3.2
Stopping Distance at High Speed

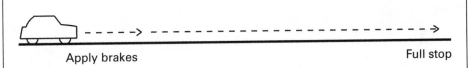

Apply brakes   Full stop

As cars began to travel at faster speeds, engineers began to worry much more about safety. An important need was to design highways with sight distances that would allow drivers traveling the speed limit to bring a car safely to a stop.

Highway engineers in New Jersey determined in the 1930s that motor vehicles with four-wheel brakes decelerate on a level road at a rate of 17.4 feet per second for every second of elapsed time. With this number, engineers could employ a formula for calculating the stopping distance required on a level road:

$$S = (1.075) \ V^2 \ / \ 17.4$$

where S is the distance in feet required to stop the vehicle and V is the speed in miles per hour.[a] At a speed of 60 miles per hour, the stopping distance was 222 feet:

$$S = (1.075) \ (60)^2 \ / \ 17.4$$

$$S = 222 \ feet$$

There was a lag of between 0.75 and 1.5 seconds from the time a driver recognized the need to decelerate and the time the driver began to brake the vehicle. An increment added to the stopping distance included this lag. Engineers designed stopping distances on the assumption that a driver would see 4 feet above the road.

The sight distances necessary for a driver to pass another car, when the other car was traveling the speed limit, required at least three times the length of the stopping distance. Solid double lines in the middle of paved American roads (or signs on the roadside) now indicate where passing is not safe or not allowed.

Automobiles were not designed for safety of occupants at high speeds until after 1966, when seat belts and other modifications became mandatory.[b] Safety features on highways also improved in the last third of the twentieth century.

Sources: Charles M. Noble, "The Modern Express Highway," *ASCE Transactions* 104 (1939): 1074. *Notes:* (a) The number 1.075 in the formula appears to have included a factor to convert feet per second into miles per hour in *V*. (b) See the National Traffic and Motor Vehicle Safety Act, Public Law 89–563 (1966).

Figure 3.8
Express highway network proposed by the Bureau of Public Roads in 1939. Courtesy of U.S. Federal Highway Administration.

Roads, proposing express highways across the United States that would extend a total of 14,336 miles (23,071 km) (figure 3.8). However, the Bureau concluded that there would be insufficient traffic to finance these highways from tolls.[24]

The Interstate Highway System

During World War II, America's roads proved strong enough to carry military traffic, but deferred maintenance caused them to deteriorate by the end of the war in 1945.[25] Anticipating a postwar need for new employment, President Roosevelt in 1941 asked a committee of experts chaired by MacDonald to produce a new postwar plan for express highways in rural as well as urban areas. The report, *Interregional Highways*, appeared in 1944 and called for a network of 33,920 miles (54,588 km).[26] The report informed the Federal-Aid Highway Act of 1944, although Republicans changed the name of the proposed highways from "interregional" to "interstate." Coming into effect at war's end, however, the highway act provided only matching funds to the cash-strapped states. During the 1930s, the number of automobile registrations had risen from 23 to 27 million; from 1946 to 1950, pent-up demand for cars increased the number of registrations to 40 million. The mileage of paved roads barely rose over the same period from 170,000 to 220,000 miles (273,588 to 354,055 km).[27]

General Dwight D. Eisenhower became President of the United States in 1953 and made express highways a priority of his administration. Having grown up in Abilene, Kansas, with no paved roads, Eisenhower appreciated the difference they could make to rural areas. Graduating from the U.S. Military Academy at West Point in 1904, he had helped to lead a convoy of army trucks across the United States in 1919, which encountered primitive roads and crossings and proved the need for better highways (figure 3.9). During World War II, as supreme commander of Allied forces in Western Europe, Eisenhower depended on roads to supply his armies, and in Germany he encountered the *Autobahnen*, express highways built in the 1930s to carry vehicles at high speeds.[28]

MacDonald retired in 1953, and the following year Eisenhower proposed to the nation's state governors a massive program to build new highways across the country. Two years later, Congress passed and the President signed the National Interstate and Defense Highways Act, providing funds over the next three years to begin construction of 41,000 miles (69,983 km) of express highways (figure 3.10). Construction of the network was largely finished by 1980. The federal government paid for 90 percent of the construction cost and the states paid the rest. A federal highway trust fund supplied by new federal taxes on the retail sale of motor fuel and by other fees met the costs of construction and further maintenance. Instead of designating existing roads, the program built mostly new highways, although many existing roads connected parts of the network. The new highways were to be free of tolls, except where an interstate route followed an existing highway that had them.[29]

Construction of the Interstate Highway System, as it came to be known, implemented standards of express highway design that had evolved in the 1930s. Each Interstate Highway was to have at least two lanes of traffic in each direction, with a median strip or divider separating traffic going in opposite directions. The Interstates would go over or under intersecting roads and rail lines, eliminating the need for stops, and access ramps and roads would allow traffic to enter and depart safely (figure 3.11). However, the American Association of State Highway Officials (AASHO) decided to conduct a series of full-scale field tests from 1958 to 1960 to confirm or improve the key standards for pavements and overpass bridges.

Another fine example of modern engineering

Figure 3.9
Transcontinental U.S. Army convoy in 1919 stopped before a bridge. Caption (satirically) reads "Another fine example of modern engineering." Courtesy of Dwight D. Eisenhower Presidential Library.

At a test site near Ottawa, Illinois, AASHO built six roadway loops. On the four main ones, different thicknesses of asphalt concrete covered half of each loop and different thicknesses of Portland cement concrete covered the other half. Small bridges, typical of highway overpasses, were built into the last two loops to test steel and concrete support beams (box 3.3). Heavy trucks were the kinds of vehicles that caused highways to wear out, and the U.S. Army supplied drivers who took turns driving trucks of varying sizes and loads around each track, day and night, for two years. The AASHO tests found that certain formulas could predict (to an acceptable degree) the behavior of different pavements under a range of truckloads. The bridges performed very closely to expectations. Understanding of roads was much better now than it had been when MacDonald began his work with the Bureau of Public Roads, and field testing was now the key to useful knowledge.[30]

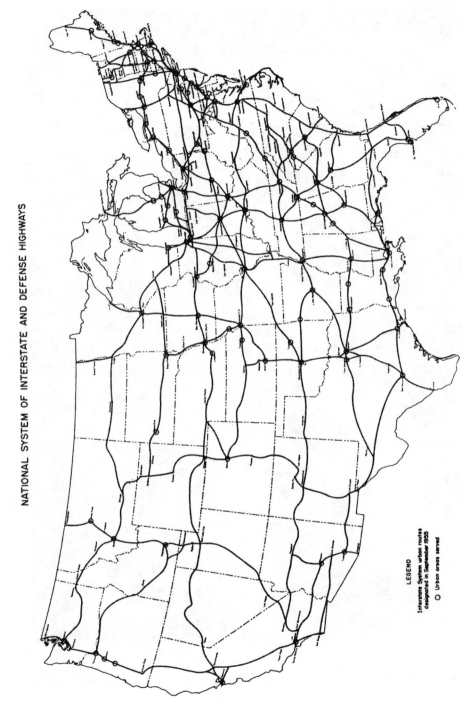

NATIONAL SYSTEM OF INTERSTATE AND DEFENSE HIGHWAYS

LEGEND

Interstate System urban routes
designated in September 1955

◯ Urban areas served

Figure 3.10

Map of Interstate Highways proposed in 1955. Courtesy of U.S. Federal Highway Administration.

Figure 3.11
Aerial view of Interstate 95 in Virginia, showing a landscaped median strip, access roads on each side, and an overpass bridge. Courtesy of U.S. Federal Highway Administration.

Box 3.3

The AASHO Road Tests

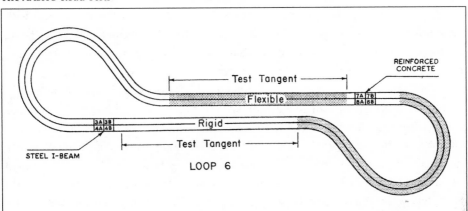

In 1958, at a test site near Ottawa, Illinois, the American Association of State Highway Officials (AASHO)[a] built six loop-shaped test roads, each divided into two kinds of pavement: relatively "flexible" sections consisting of asphalt roadway (shaded) and "rigid" sections consisting of concrete made with Portland cement (unshaded), with test bridge crossings on the last two loops. On each two-lane track, U.S. Army drivers took turns driving trucks of varying weight. The trucks drove day and night for two years.

To calculate the stress in pavement (reinforcement along the edges was a separate concern), engineers can use a formula, $S = (2.4) W / d^2$, where W is the axle load in kilopounds, or kips (1 kip = 1,000 lbs), d is the thickness of the concrete (in inches), and S is the stress in the concrete in pounds per square inch (psi).

Concrete will crack under a stress of 3,000 psi. On a 6-inch-deep pavement, a heavy truckload of 69 kips would exceed the breaking stress of the concrete:

$$S = (2.4) W / d^2$$

$$S = (2.4) \, 69{,}000 / 6^2$$

$$S = 4{,}600 \text{ psi}$$

On an 8-inch pavement, however, the stress would come to 2,587 psi, safely below the 3,000-psi limit. Engineers designed the interstate highways to have pavement depths able to support loads typically carried by heavy trucks.

Source: Highway Research Board, *The AASHO Road Test*, 7 vols. (Washington, DC: National Academy of Sciences—National Research Council, 1961–1962), vol. 1, pp. 19 and 25. Formula is from the American Concrete Pavement Association (ACPA). *Note:* (a) Now the American Association of State Highway and Transportation Officials (AASHTO).

Interstate Highways came to be numbered like U.S. Highways but with the starting points reversed, so that odd-numbered Interstate routes going north to south began with low-numbered routes on the West Coast, such as Interstate 5, or I-5, and with high-numbered ones along the East Coast, such as I-95. Even-numbered routes going east to west began with I-10 along the southern boundary of the United States, and ended with I-90 and I-94 near the Canadian border. The U.S. Highways had a distinctive white shield; Interstate Highways had a different shield colored red, white, and blue.[31]

The new thoroughfares included expressways through and around major cities, as well as across open country, helping to make America a more interconnected civilization. After World War II, millions of veterans received allowances and loans from the federal government to further their education and buy houses in the suburbs. David Bohannon in California, followed by builders such as William Levitt in New York and Pennsylvania, built affordable housing for them (figure 3.12). Schools and shopping centers spread with the new housing and new highways.[32]

In 1950, 40 million cars served a U.S. population of 151 million. By 1980, the transition to an automotive society was essentially complete, as 121 million cars served a population of 226 million; 20 years later, these numbers had risen only to 133 million cars in a population of 281 million. The total length of asphalt and concrete roads reached 1,032,000 (1.6 million km) in 1980, with only 316,000 more miles (508,552 km) added by 1995. The length of Interstate Highways rose to about 42,000 miles (67,592 km) and unsurfaced roads fell to about one-fifth of all roads.[33]

In 1920, half of the American people lived in communities of 2,500 or more, but only about one-third lived in the larger concentrations of cities and suburbs that would today be called metropolitan areas. By 1950, more than half of all Americans lived in the latter areas. At the end of the twentieth century, half of the American people lived in suburbs and 80 percent lived in metropolitan areas. Affordable cars and new roads and highways made the postwar migration to the suburbs possible.[34]

Controversy attended this massive change. In the 1930s, less-populated states protested federal highway policy that favored cities. Twenty years later, as new urban expressways disrupted older neighborhoods, people in the cities protested. The loss of middle-income people to the suburbs lessened the ability of cities to maintain public services for the lower-income residents who remained in them. Railways lost their preeminent role in transportation as many people chose the greater flexibility and freedom of motor vehicles and roads to the fixed lines and timetables of the railroads. This freedom brought with it a dependence on motor vehicles, not only for personal travel but for food and other goods shipped to retail shopping locations that in the suburbs required cars to reach.[35]

Figure 3.12
Postwar houses in Levittown, Long Island, New York. Courtesy of U.S. Library of Congress.

Automobile safety became a concern when Ralph Nader wrote *Unsafe at Any Speed* (1965), a book exposing dangers in the design of cars at the time. Federal law soon required the inclusion of safety belts and shatter-resistant windshields.[36] Postwar America also took low oil prices for granted. When the price of oil rose in the 1970s, Japanese car makers introduced smaller and more fuel-efficient vehicles. Only after losing one-third of the domestic automobile market did American firms begin to offer more energy-efficient cars of their own. New federal and state regulations reduced the most toxic emissions from motor fuels, but the use of fossil fuels continued and added carbon dioxide to the atmosphere. In urban areas, motor vehicle traffic grew and brought congestion that filled many of the new roads and highways. These effects challenged modern life as America and the world moved into the twenty-first century.[37]

Highways and Crossings: Marilyn Reece and Arvid Grant

The new highways could not have functioned without bridges and overpasses. Most bridge and overpass structures were steel or concrete roadway spans between concrete abutments, many with one or more concrete pillars underneath. Many of these were built in the postwar era quickly to keep pace with highway construction, and most followed standard designs. These bridges were serviceable at the time they were built, but unnoticed design flaws in some and deterioration of many more over time reduced margins of safety, causing a few to fail. Design standards are much better now, although a backlog remains of older bridges that the public needs to repair or replace.[38]

Interchanges are the places where two express highways intersect, and standard interchanges typically had a circular loop pattern to accommodate turns (figure 3.13).

Figure 3.13
Aerial view of California Route 110 and Interstate 10 interchange in Los Angeles in 1963, showing a conventional loop design. Courtesy of U.S. Federal Highway Administration.

Figure 3.14
Marilyn Jorgenson Reece supervising construction of the Interstate 10 and Interstate 405 interchange in Los Angeles. Courtesy of California Department of Transportation (Caltrans).

Marilyn Jorgenson Reece, the first woman to be a licensed civil engineer in California, improved the design of interchanges (figure 3.14). For a new interchange in Los Angeles where Interstate 10 and Interstate 405 met, Reece swept out each of its four curved turns more gradually (figure 3.15). The result was an interchange, completed in 1964, that set an example of efficient and economical design that was also safer for drivers and more elegant.[39]

Crossing major rivers and straits required larger bridges. Before World War II, the longest spans in the United States were cable suspension bridges. The George Washington Bridge over the Hudson River in New York, completed in 1931, had a center span of 3,500 feet (1,066 m); and the Golden Gate Bridge across the entrance to San Francisco Bay was finished in 1937 with a center span of 4,200 feet (1,280 m).[40]

Figure 3.15
Aerial view of Interstate 10 and Interstate 405 interchange in Los Angeles in 1964, showing the swept-out design of Reece. Courtesy of California Department of Transportation (Caltrans).

After World War II, American engineers began to experiment with a different kind of bridge, using cable stays, which were practical over shorter crossings. In a cable-stayed bridge, straight cables extended from one or more post-like towers to hold up a roadway deck. Designed by the civil engineer Arvid Grant (1920–2014), the East Huntington Bridge over the Ohio River, between West Virginia and Ohio, had cable stays to carry two lanes of traffic across a 900-foot span (figure 3.16). Completed in 1985, the bridge won praise for its graceful appearance. Cable-stayed bridge spans continued to be built into the twenty-first century.[41]

Figure 3.16
The East Huntington Bridge over the Ohio River. Courtesy of Arvid Grant and David Goodyear.

The Rising Skyline: Chicago and New York

As modern civilization in America spread out into suburbs and rural areas, the buildings at the center of larger cities rose dramatically higher. The invention of mechanical elevators or lifts in the 1860s had made it practical for people to occupy buildings higher than three or four floors, and by the twentieth century urban land was more expensive as a result. Developers began to commission taller buildings, mainly to accommodate business offices. As these buildings rose in height, they came to be known as skyscrapers. The key to their engineering was a structural framework composed of interconnected vertical columns and horizontal beams made of modern iron or steel.

Started in 1891, the Monadnock Building in Chicago rose to sixteen floors (figure 3.17). Designed by Daniel Burnham and John Wellborn Root, the building began as two identical adjoining structures. Heavy brick walls carried the weight of the floors but an iron frame gave the walls some reinforcement. An additional two buildings adjoining to the south completed the structure in 1893. The last of these two additions

Figure 3.17
The Monadnock Building in Chicago, in the background, viewed from the north. Courtesy of U.S. Library of Congress. Cervin Robinson, photographer.

had a steel frame to carry its weight. The Monadnock Building (at 53 West Jackson Boulevard) covered a city block and was the largest office building in the world at the time of its completion. The steel frame of its last section supported less heavy masonry "curtain" walls, so-called because they no longer had to carry the building's weight and simply held windows and kept out the weather. Buildings with steel weight-bearing frames and curtain walls became the standard in subsequent skyscraper design.[42]

In the early 1900s, New York City began to erect taller structures. The Woolworth Building, finished in 1913, consisted of a building with 30 floors topped by a narrower tower having an additional 30 floors. Tall buildings in New York rose higher during the 1920s with the Chrysler Building, finished in 1930 with 77 floors. The Empire State Building, with 102 floors, was the tallest building in the world at its completion in 1931 (figure 3.18).[43]

Fazlur Khan and the Revolution in Height

After 1945, real estate in the central business districts of the largest cities in America continued to rise in value, and demand for new office space revived the construction of tall buildings. These new structures were expensive to build higher than 40 floors, though, because of wind forces that increased with height. Finding a way to build taller structures with much greater efficiency and economy was the work of Fazlur Rahman Khan (1929–1982), an immigrant civil engineer from what is now the nation of Bangladesh (figure 3.19).

Khan belonged to a family of educated Muslims in the city of Dhaka. He attended an engineering college in Kolkata in British-ruled India but had not finished his studies in 1948 when the subcontinent divided into the independent nations of India and Pakistan. Dhaka and the surrounding predominantly Muslim region became part of Pakistan (and achieved independence as Bangladesh in 1971). Riots attended the 1948 partition, and for his safety Fazlur had to return home. He admired the action of his father, the head of a local college in Dhaka and a noted religious moderate, to provide refuge to Hindu students and stand down the threats of extremists. Fazlur Khan completed a degree in civil engineering in Dhaka and won a Fulbright scholarship to study at the University of Illinois in the United States. After returning to work in east Pakistan, his skills were not fully employed, and in 1960 he was able to find work in Chicago with the building design firm of Skidmore, Owings & Merrill. Here he began to rethink the design of tall buildings.[44]

The steel framework of a tall building consists of vertical columns connected by horizontal beams that support each floor. The weight of the building exerts a compressive

Figure 3.18
The Empire State Building in New York. Courtesy of U.S. Library of Congress. Carol M. Highsmith, photographer.

Figure 3.19
Fazlur Rahman Khan. Courtesy of *Engineering News-Record* (February 10, 1972).

force in the columns, which must also resist horizontal wind forces and any move-ment of the foundations. Early twentieth-century steel skyscrapers consisted of closely spaced columns connected by cross-beams, with masonry curtain walls to cover the steel framework. Although less massive than walls that had to bear weight, these walls were still heavy and reduced the amount of window space. During the 1950s and 1960s, architects and engineers designed glass and steel buildings that were less massive, but above 40 floors, the cost per square foot of floor space increased because of the need to withstand wind forces that increased with altitude.

Khan thought deeply about this cost in the early 1960s and solved it with an inno-vation that came to be known as the *framed tube*. A cardboard box had strength against bending, and Khan realized that if a tall building placed most of its strength in its out-side columns and beams, it would form a box with greater stiffness against wind than a building in which the strength was distributed more equally among the columns and beams outside and inside. If this insight was correct, then a building could use less steel inside and still be safe at higher elevations.[45]

Khan believed that a building with a stronger outer structure in the form of a square or rectangular tube could rise to 100 floors and be economical. He tested a framed tube in a 43-floor concrete apartment building in Chicago, the DeWitt–Chestnut Building. Then in 1965 he designed a glass and steel tower in Chicago, the John Hancock Center, as a framed tube (box 3.4) (figure 3.20). On its completion in 1969, the Hancock tower rose 100 floors to an altitude of just over 1,100 feet (335 m) and used 29.7 pounds of steel per square foot of usable floor area. The Empire State Building, with 102 floors and almost the same area, used 42.2 pounds of steel per square foot. In addition to its tubu-lar design, the Hancock tower featured cross-bracing on its outer walls, which allowed the columns to be spaced farther apart and to receive added strength. The framed tube soon became the method of designing very tall buildings in the United States and around the world, although only some buildings added exterior cross-bracing.[46]

The Hancock tower was a mixed-use building, with offices and shopping on the lower floors and apartments above them. Buildings of this kind helped to attract some younger people back to the cities in the late twentieth century.[47] Khan collaborated with an architect, Bruce Graham, but in their work together, Khan gave expression to his own aesthetic vision as well as his own structural insight.[48] The Hancock tower was also easy for the public to understand: the visible exterior columns, beams, and bracing showed how the structure held together. In 2018, the name of the tower changed to its address, 875 North Michigan Avenue.

Fazlur Khan's next great building was his tallest, the Sears Tower in Chicago (now the Willis Tower) (figure 3.21). Khan designed the structure in the form of nine tubes

Box 3.4

The John Hancock Center (now 875 North Michigan Avenue)

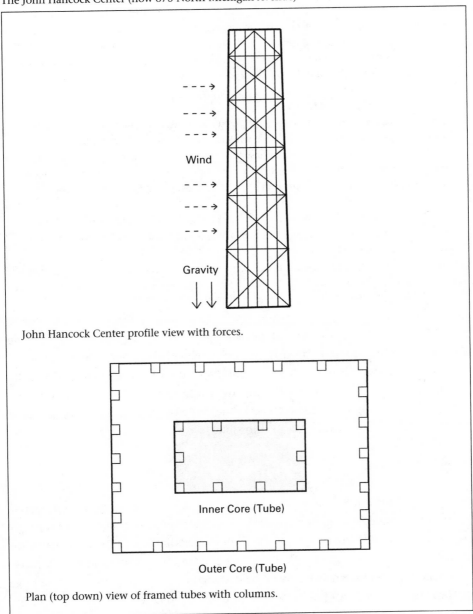

John Hancock Center profile view with forces.

Plan (top down) view of framed tubes with columns.

Box 3.4 (continued)

Fazlur Khan designed the John Hancock Center tower to be a framed tube, in which a rectangle of columns on the outside formed an outer core or tube, and another rectangle of columns inside formed an inner core, with diagonal braces added to connect the outer columns. Each vertical column needed enough steel so that the stress from the loads it had to withstand did not exceed the *allowable stress* of the steel, the amount of stress the steel could safely carry. For the Hancock tower, the allowable stress of the steel is 29,000 pounds, or 29 kips, per square inch (29 ksi).

Stress comes from the vertical force of gravity and the horizontal force of wind. The inner tube carries half of the gravity load, while the outer tube carries the other half of the gravity load and all of the wind load. These loads are greatest at the base. The base of each steel column in the outer tube carries stress from gravity of 19 ksi. The wind stress at the base of each column is 9.5 ksi. The total stress of 28.5 ksi is therefore just below the allowable stress of the steel.[a]

The efficiency of the building is the total stress divided by the allowable stress, 28.5 ksi / 29 ksi, or 98.3 percent. The building saved steel by meeting most of the total load of gravity and wind with the outside columns.

Source: David P. Billington and graduate students, *Structural Studies* (Princeton University, 1984–). *Note:* (a) Allowable stress in steel is normally one-half of the breaking stress; thus, the building was quite safe.

bundled together. Two of the tubes stopped at the 50th floor, another two reached the 66th floor, and three more stopped at the 90th, leaving two more to rise to 110 floors, or 1,450 feet. The lower tubes reinforced the upper ones in an efficient and visually unique way. At its completion in 1974, the tower was the tallest building in the world, and it still has the highest occupied floor in the United States. With a bundled tube of this kind, even taller structures became possible.

Machine and Structure

The automobile was the dominant machine of the twentieth century. Mass-produced and mostly privately owned, the car was and is a means to personal freedom through the mobility it affords. But cars are useless without roads and bridges, most of which are publicly owned. These and other built structures all are fixed in place. They are made with manufactured materials but must be designed to serve local needs and settings that are unique. Perceptions of modern technology often reflect a machine-like image of uniform design, use everywhere, and rapid obsolescence. But there is a different side of technology, structure, that when done well stands for and respects another set of ideas: the unique, the local, and the enduring.

Figure 3.20
The John Hancock Center (now 875 North Michigan Avenue), Chicago. Courtesy of J. Wayman Williams, photographer.

Figure 3.21
The Willis (former Sears) Tower in Chicago. Courtesy of U.S. Library of Congress. Carol High-smith, photographer.

A more sustainable future will require not just new technology but a better balance within technology between the ideas of machine and structure. Machines are mass-produced, mostly for private use, and are rapidly disposable. To give more attention to the unique needs of place, which are often public needs, and to the need for sustainability, is to give more emphasis to ideas associated with structure.[49]

After 1945, public engineers achieved three major new innovations: nuclear power, the jet engine, and rockets capable of sending astronauts into outer space. Nuclear reactors generated power in a radically new way, and the jet engine and the rocket revolutionized the ability to fly higher and faster than ever before.

The great federal infrastructure projects of the twentieth century brought new water control, electric power generation, and new roads and highways to the American landscape. After 1945, the federal government achieved three new engineering break-throughs: unlocking the energy in the atom, reaching unprecedented speeds in the air with the jet engine, and, finally, sending astronauts to the Moon. The first of these was the use of nuclear energy, as an explosive and then as a controlled source of power.

During World War II, in a program run by the U.S. Army, the United States invented bombs that released explosive energy through nuclear fission. Two of these brought the surrender of Japan in the summer of 1945. In the postwar years, Captain Hyman Rickover of the U.S. Navy engineered the controlled use of nuclear fission to propel submarines. Rickover faced opposition within the navy to this innovation and he needed the political support of Congress to prevent his forced retirement in the early 1950s. The successful 1954 voyage of his first nuclear-powered submarine, the USS *Nautilus*, and its trip to the North Pole in 1958 under the Arctic icepack, showed the practicality of nuclear-powered ships, and the navy added more of them.

During the 1950s and 1960s, America began to generate electricity for civilian use through nuclear power plants on land, most of them based on Rickover's engineering. Advocates believed that nuclear energy would be more economical than fossil fuels. Nuclear plants proved expensive to build, though, and a reactor breakdown at the Three Mile Island plant in Pennsylvania in 1979 placed new plant construction on a hold that has proved enduring. America continues to receive one-fifth of its electricity from nuclear power plants, which work without emitting carbon dioxide into the atmosphere as fossil fuel plants do. However, the public remains divided over the risks of nuclear power. An alternative source of nuclear energy, from fusion, has been a focus of research since the 1950s, but the technology has not yet sustained the generation of more energy than it consumes. Rickover recognized that the key to obtaining power from nuclear fission was successful engineering, and he saw that an engineered energy system that carried great risks required great discipline to manage safely.

Nuclear Energy and the Atomic Bomb

In chemical reactions, such as the combustion of oil, molecules break apart or merge as the atoms rearrange themselves into new molecules, releasing energy in the form of heat. In the early twentieth century, scientists discovered that the nuclei of atoms could also break apart or merge and release energy as heat. Reactions in which the nuclei broke apart came to be known as nuclear *fission*, while those in which they merged came to be called nuclear *fusion*. In 1905, Albert Einstein formulated the relationship $E = mc^2$, in which the amount of energy (E) equals the mass (m) of a body of atoms multiplied by the speed of light (c) squared. Scientists calculated that a nuclear fission reaction could release several million times the energy of a chemical reaction.[1]

During the 1930s, it became clear to a handful of scientists in Europe that nuclear fission might be possible to engineer to create explosives. Scientists aware of this possibility were able to take refuge in the United States before World War II. One of them, Albert Einstein, signed a letter to President Roosevelt in August 1939 warning that enemies of the United States might develop and use nuclear weapons unless America acquired them first. After America's entry into World War II on December 7, 1941, the U.S. Army Corps of Engineers organized a program, code-named the "Manhattan Engineering District," to develop nuclear bombs.[2]

Nuclear weapons were most practical to make from atoms of the element uranium. Most atoms have one or more electrons around a nucleus that contains an equal number of protons. Atomic nuclei also contain one or more neutrons, except for the simplest form of hydrogen, which has none. Fission occurs when a free neutron penetrates an atomic nucleus having protons and neutrons, causing the nucleus to break apart and release energy. Heavier atoms are easier to break apart, and the Manhattan Project began its work with uranium because it had the heaviest nuclei. But the project needed a particular form of uranium to achieve fission.

Nuclei that have the same number of protons can occasionally vary in the number of neutrons they contain. When comparing atoms of an element that differ in this way, the atoms are called isotopes of the element. Most atoms in natural uranium have the atomic number 238, which is the sum of the 92 protons and 146 neutrons in the nucleus. These atoms are also known as uranium-238 isotopes. About 0.7 percent of the atoms in natural uranium have 92 protons and only 143 neutrons and are known as uranium-235 isotopes. These have less stable nuclei and are the easiest to fission.

When an outside neutron fissons a nucleus of uranium-235, the fission ejects neutrons that can fission other uranium-235 isotopes nearby. Scientists believed that natural uranium enriched to about 3 percent uranium-235 would fission at a constant rate, producing steady heat. A concentration of 70–90 percent uranium-235 would fission at an accelerating rate, leading within a fraction of a second to an enormous explosion. Such high concentrations of uranium-235 would be necessary to make nuclear weapons. Controlled fission also caused small amounts of uranium-238 to absorb a neutron and change into plutonium-239, which was also highly fissile. The Manhattan Project aimed to engineer bombs made from uranium and plutonium.

The first need was to produce a controlled reaction. At the University of Chicago in December 1942, scientists placed samples of natural uranium alongside blocks of graphite in ascending layers (figure 4.1). The uranium naturally released a small number of neutrons, and a fluid or a solid material, known as a *moderator*, could slow this release. This slowing was necessary to fission uranium in a controlled way, and graphite was a good moderator. The withdrawal of cadmium control rods, which absorbed free neutrons, allowed the slowed neutrons to fission small amounts of uranium-235 in the test samples. The fission created a limited amount of heat at a constant rate. Reinserting the control rods ended the fission. The Chicago demonstration confirmed the possibility of a controllable reaction and verified certain calculations about the rate of fission.[3]

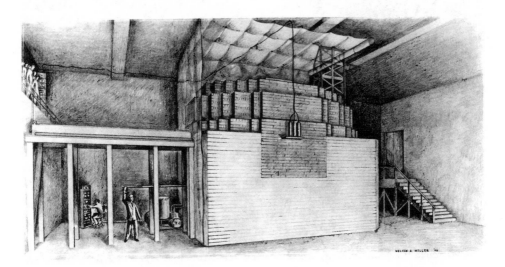

Figure 4.1.
Drawing of the graphite reactor ("Chicago Pile I") at the University of Chicago, 1942. Courtesy of Argonne National Laboratory.

A bomb would require an accelerating reaction. At Oak Ridge, Tennessee, the army built a facility to increase or "enrich" the amount of uranium-235 in natural uranium. The process of enrichment began by mixing powdered natural uranium with a chemical to produce a gas. After some processing, the gas was spun in centrifuges that left a residue containing a higher concentration of uranium-235; repeated spinning raised the amount of uranium-235 in the residue. The enrichment process at Oak Ridge demanded enormous quantities of electricity (in the form of direct current), and to supply it, the TVA built Fontana Dam nearby. At Hanford, Oregon, a plant with several graphite reactors made plutonium-239, which was also possible to accumulate.

By 1945, both the Hanford and Oak Ridge plants had produced enough fissionable material for one uranium bomb and two plutonium bombs. These received final assembly in a laboratory at Los Alamos, New Mexico. The plutonium bombs had a more complex design and as a test the laboratory detonated one successfully in July in the New Mexico desert (figure 4.2). The dropping of a uranium bomb on Hiroshima, Japan, and

Figure 4.2
The test explosion of a plutonium nuclear bomb in the New Mexico desert on July 16, 1945. Courtesy of Los Alamos National Laboratory.

the dropping of the remaining plutonium bomb on Nagasaki, both in August 1945, ended World War II.[4]

A year later, the Truman administration converted the Manhattan Project into a new civilian agency, the Atomic Energy Commission (AEC), under the former TVA chairman, David Lilienthal. But the principal mission of the AEC was to build nuclear weapons for the U.S. armed forces as the Cold War with the Soviet Union intensified in the late 1940s.[5] The control of nuclear energy for use as a source of power would not be the work of a civilian agency but of a military service, the United States Navy.

Rickover and Nuclear Power

The U.S. Navy did not have a wartime role in nuclear energy. After the war, the efforts of a remarkable officer gave it one. Born in Poland to Jewish parents, Hyman Rickover (1900–1986) emigrated to Chicago, Illinois, at the age of six and later worked as a telegraph messenger. An expanded need for officers in World War I enabled him to secure an appointment to the U.S. Naval Academy in 1918, which he entered after independent study to make up for deficiencies in his education. Enduring anti-Semitism from other classmates, he received a commission in 1922 and earned a reputation for careful work in his first shipboard assignments. Attracted to technical challenges, Rickover earned a master's degree in electrical engineering at Columbia University in 1929, served on submarines for several years, and transferred to full-time engineering duty in 1937. Engineering duty was a less prestigious career track that meant assignment on shore. In December 1940, though, he took charge of electrical engineering for the navy's Bureau of Ships, which designed the nation's warships and oversaw their construction. Rickover held the electrical position for the duration of World War II, ending the war with the rank of captain.[6]

In June 1946, the Bureau of Ships sent a small group of officers with engineering experience, including Rickover, to Oak Ridge, where they joined engineers from the army and private industry to study the practicality of generating electric power from nuclear energy. Rickover quickly realized how useful nuclear energy could be to submarines. To travel, submarines used internal combustion engines on the ocean surface and electric batteries when submerged, requiring the ship to surface often to recharge the batteries and replace fuel oil. If nuclear energy from a small amount of uranium could be employed to heat water into steam, the steam could be used to power engines, allowing submarines to operate submerged for weeks or months at a time.

The technical difficulties of using nuclear energy were also clear to Rickover, starting with the fact that a practical system to convert fission energy into electric power did not exist. A system for a submarine would be especially challenging. Space would be confined and a nuclear propulsion system would have to absorb the shock of motion and combat. There was also the special danger of radiation. Single neutrons emitted in the breakdown of atomic nuclei were not harmful to humans, but as nuclei broke apart, they released clusters of protons and neutrons that were dangerous. Fission also emitted the highest frequencies on the spectrum of electromagnetic radiation, gamma rays, and a short period of exposure to these could be fatal. Humans onboard would need to be shielded from these radioactive effects at all times.[7]

Rickover mastered what was known of nuclear engineering at the time and took charge of the naval group at Oak Ridge, whose members resolved that the future of the navy lay in nuclear propulsion. However, the AEC had control over nuclear technology and its priority was to make bombs. The navy was interested enough to have sent Rickover and his group to study nuclear energy but did not see an urgent need to develop it for use to power ships. Rickover's Jewish origin also stood in his way, as did a caustic side of his personality. Although himself respected for his efficiency, he often expressed disapproval of others' failings. The Bureau of Ships dissolved Rickover's nuclear group in the fall of 1947 and assigned him to an office in a converted restroom in Washington, hoping perhaps that he would leave the service.[8]

However, with the help of supportive and socially connected younger officers, who also saw the value of nuclear propulsion, Rickover secured the approval of Fleet Admiral Chester Nimitz, the chief of naval operations and a former submariner, for a program to develop a nuclear-powered submarine. In July 1948, the Bureau of Ships changed its mind and appointed Rickover to head the program. He also obtained appointment in February 1949 as naval liaison to the AEC, which made him an official there as well. In these dual capacities, Rickover was able to achieve the cooperation he needed from the civilian agency to proceed.[9]

To generate power from fission, Rickover and the engineers he recruited to his new program designed a propulsion system consisting of a nuclear reactor, a primary loop of piping, and a secondary loop of piping (figure 4.3). The reactor would be a stainless-steel containment vessel in which uranium, enriched to 3 percent uranium-235, would be immersed in a liquid that would serve as a moderator. The enrichment to 3 percent was enough to sustain a controlled reaction that would heat the surrounding fluid to a steady temperature. The primary loop would carry the heated fluid to a container where it would heat water in the secondary loop into steam. The steam would then turn the blades of a horizontal turbine shaft, which would rotate the propellers on a submarine and also generate electricity for use onboard. The fluids in both loops would then be cooled and the cycle would repeat itself. The reactor and the fluid in

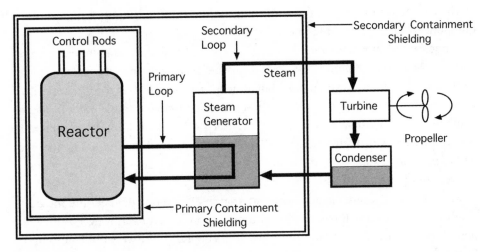

Figure 4.3
Pressurized water reactor and propulsion system. Courtesy of U.S. Navy (adapted from original diagram).

the primary loop would be radioactive and had to be shielded from the water in the secondary loop and from the rest of the submarine and its crew.[10]

Rickover decided to develop two kinds of reactors. In the first, a pressurized water reactor, the moderator and fluid in the primary loop would be ordinary water, kept under pressure so that it could reach a temperature of 600°F (315°C), above the boiling point of unpressurized water of 212°F (100°C). In the second kind of reactor, liquid sodium would be the moderator and primary fluid. Fluid sodium needed a higher temperature than water but not as much pressure. Westinghouse agreed to make a pressurized water reactor for Rickover at a laboratory near Pittsburgh, Pennsylvania, while General Electric agreed to make a sodium reactor at a laboratory near Schenectady, New York. The pressurized water reactor would go into the first nuclear submarine, the USS *Nautilus*. Sodium proved highly corrosive, though, and Rickover abandoned plans to use a sodium reactor in a second submarine, the USS *Seawolf*. He gave the *Seawolf* a second pressurized water reactor instead.[11]

The idea of a small pressurized water reactor had been sketched in 1946 in an eight-page manuscript paper by the physicist Alvin Weinberg, director of the Oak Ridge facility.[12] But no one at Oak Ridge gave further thought to engineering such a reactor. One of Rickover's colleagues recalled the emphasis at the time: "Most of the focus in any reactor discussion was reactor physics, which in turn relied mostly on sophisticated mathematical techniques. . . . But with characteristic bluntness Rickover had a simpler question: 'What are you going to make the damn thing out of?'"[13]

A nuclear fission reactor and its piping would need to withstand high temperatures and resist corrosion. Stainless steel could provide acceptable material for the reactor containment vessel walls and piping. But enriched uranium fuel, in the form of pellets, would need to be stacked in long holders or rods and inserted into the water inside the containment vessel for a reaction to sustain itself. Stainless steel, if used in such rods, would interfere with fission because stainless steel absorbed neutrons. The fuel rods had to be strong enough to withstand heat and corrosion inside the containment vessel, yet not interfere with fission.

Rickover found a rare metal, zirconium, that in alloy form could endure the harsh conditions in the containment vessel and not absorb neutrons. However, only a few pounds of zirconium existed in the United States because it was very difficult to make. Rickover promptly ordered several tons. Naval practice was to let contracts to private suppliers, who after a year's delay failed to deliver. Finally, Rickover was able to seize control of the zirconium order and began making it himself in a special facility. To work as required, the zirconium needed to have the element hafnium removed from it, and this element fortunately answered another need, which was a way to regulate the fission of neutrons in the reactor. Hafnium absorbed neutrons, and inserting or removing separate hafnium control rods made it possible to adjust the rate of fission in the reactor more precisely and to shut down the reaction if necessary.[14]

Building the USS *Nautilus*

With many engineering questions still to be answered, in the summer of 1950 Rickover ordered construction of a full-scale test version of a propulsion system, the "Mark I," at a remote facility in Idaho now known as the Idaho National Laboratory. Rickover designed the test version to be enclosed in a steel hull and then submerged in a giant tub of water (figure 4.4). "The engineers howled in unison at the idea," wrote one of them. The usual way to build a complex and unfamiliar assemblage of machinery was to build each part first and test all of them separately in a large building with plenty of room. "We don't have time to do it that way," replied Rickover.[15] The captain built his giant tub and submerged in it a full-sized submarine compartment with a Westinghouse reactor and turbine system inside. He trained a group of handpicked officers and enlisted sailors to operate and repair the complex equipment, instructed them in the engineering and science they needed to know, and emphasized safety and rigorous adherence to procedures.

Figure 4.4
The Mark I test system. Reactor compartment submerged in a tub. Courtesy of Idaho National
Laboratory.

On March 30, 1953, the Mark I generated a controlled fission reaction that produced heat. In June, the test system was able to operate its turbine safely at full power for 48 hours. Rickover horrified his engineers again by ordering the system to keep going for a total of 96 hours to try to discover any flaws before the system went onboard a submarine. The system passed this test. When the navy's own shipyard in Maine declined to build a submarine to contain the new system, Rickover found a private shipyard in Connecticut that needed the contract, the Electric Boat Division of General Dynamics. He ordered completion by January 1, 1955.[16]

In 1951 Rickover was up for promotion from captain to rear admiral. He remained unpopular with many senior officers in the navy, and the board charged with making promotions did not recommend him. Officers passed over for promotion had to retire after 31 years, and the following year would be his last. A year later, in June 1952, the board again refused to promote him. Rickover's younger subordinates again swung into action. The nuclear program had advanced with phenomenal speed under his direction, and no one else yet had the knowledge and capacity to replace him. In testimony before Congress, Rickover had impressed powerful members of the House and Senate, and, alerted by Rickover's colleagues, several senators and representatives expressed alarm at the impending loss of the nuclear program's driving force. With the help of a sympathetic *Time* magazine correspondent, who blasted the navy for anti-Semitism and hidebound practices, a public outcry caused Congress to hold up all naval promotions. The Eisenhower administration reviewed promotions after it took office in early 1953 and Rickover advanced to the rank of rear admiral (figure 4.5).[17]

By the end of 1954, the new submarine was ready. The USS *Nautilus* had the same name as the fictional submarine in the 1870 Jules Verne novel *Twenty-Thousand Leagues Under the Sea*; an experimental submarine built by the American Robert Fulton in 1800–1801 also bore the name, as did a World War II submarine. Rickover's *Nautilus* measured about 320 feet long and 28 feet wide, and weighed about 4,000 tons (3,628 metric tons) submerged when fully loaded with supplies and a crew of about 12 officers and 90 sailors. The ship could travel underwater as well as on the surface at a speed of more than 23 knots (26 mph, or 41 km/hr) (figure 4.6). Conventional submarines could stay submerged for up to two days; the Nautilus could remain underwater for two weeks. By the 1960s, nuclear submarines could remain underwater for three months.[18]

A key feature of Rickover's program was his selection of personnel and insistence on accountability. Rickover personally interviewed officers for nuclear duty and made the interviews stressful to test an applicant's composure under pressure. Rickover also demanded that personnel assigned to nuclear duty have clear assignments so that if anything went wrong, a responsible individual could be identified. In testimony to

Figure 4.5
Rear Admiral Hyman Rickover in 1955. Courtesy of U.S. Naval History and Heritage Command.

Congress in May 1979, he observed: "Unless you can point your finger at the one person who is responsible when something goes wrong, then you have never had anyone really responsible."[19]

After sea trials, the *Nautilus* entered service in April 1955. In the months before, naval officers now realized that command of a nuclear submarine could be a career distinction and applied for the position. Rickover brushed them aside and made his own choice. In 1958, the *Nautilus* made history by becoming the first vessel to reach the North Pole by traveling under the Arctic icecap. The third nuclear submarine to

Figure 4.6
The USS *Nautilus* undergoing sea trials in January 1955. Courtesy of U.S. National Archives.

be launched after the *Nautilus*, the USS *Skate*, surfaced at the North Pole in 1959. The *Nautilus* remained in service until 1980.[20]

Rickover went on to build pressurized water reactors to power aircraft carriers and other large surface warships as well as more submarines. However, rapid expansion of the nuclear-powered fleet brought new dangers. In 1963, the nuclear submarine USS *Thresher* sank during a deep-sea dive in the North Atlantic with the loss of all on board. A naval investigation attributed the loss to a failure in the steam piping (from improper maintenance in a shipyard) that shut down the electrical system and then the nuclear reactor. Afterward, the navy instituted a program of more stringent control, not only over the entire propulsion system but over the design and assembly of all critical components on submarines. Rickover trained nuclear personnel in new steps to restore propulsion quickly in the event of an unexpected failure.[21]

Rickover eventually won promotion to full admiral and remained in charge of the navy's nuclear program until 1982, when he retired as the longest-serving officer in U.S. naval history. A corps of naval nuclear engineering personnel trained by him continued his rigorous standards, and one officer who served under Rickover from 1952 to 1953, Lieutenant Commander James Earl Carter, became 39th President of the United States, from 1977 to 1981. In a memoir, Carter wrote of the admiral: "He expected the maximum from us, but he always contributed more."[22] Rickover created a new kind of public engineering organization within the navy that achieved an impressive record of efficiency and safety after 1963. Unfortunately, the industry that developed nuclear power for civilian use had to learn on its own the need for careful control.

Civilian Nuclear Power and Three Mile Island

At first, nuclear technology was not available for civilian use. Russia's detonation of a fission bomb in August 1949, however, caused many in the United States to worry about the nation's future nuclear competitiveness. After taking office in 1953, President Eisenhower pledged his support for the peaceful use of atomic energy, and in July 1954, Congress passed an Atomic Energy Act that approved the use of nuclear reactors for civilian purposes. Insurers were still wary of the risks if anything went wrong, so in 1957 Congress passed the Price–Anderson Act, agreeing to insure utilities against nuclear power plant losses above a certain amount.[23]

Rickover offered to build a pressurized water reactor for a private utility if the utility agreed to operate and maintain it. Duquesne Light and Power of Pittsburgh accepted the offer, and at Shippingport, Pennsylvania, a village on the Ohio River a few miles north of Pittsburgh, construction began in 1954. With Westinghouse contracted to build the reactor, Rickover oversaw the project with his usual exacting attention to detail. Cost overruns still mounted as the contractors struggled with unfamiliar challenges and a tight schedule, doubling the final cost to $76 million. But finally, in May 1958, the Shippingport plant began supplying 60,000 kilowatts of power to the Pittsburgh area. The plant operated successfully until its closure in 1982.[24]

Civilian nuclear power grew from supplying less than 1 percent of the electricity generated in the United States in 1960 to 12 percent of the electric power generated in 1978. But nuclear energy was troubled. Although less expensive to operate, nuclear plants were more costly to build than coal-fired plants, and electricity generated by nuclear power proved no cheaper to consumers than electricity from burning fossil fuels. The spent fuel from nuclear reactors was highly radioactive, and no state wanted to accept it for disposal; as a result, wastes had to be stored at each plant. After the

National Environmental Policy Act of 1969 became law, utilities had to file environmental impact statements in order to build new power plants, which limited available sites, and litigation by opponents of nuclear power began to delay new construction. Most seriously, the private utilities that managed civilian nuclear power stations did not all maintain Rickover's high standards of safety and managerial control.[25]

On Three Mile Island, in the Susquehanna River near Harrisburg, Pennsylvania, local utilities built three nuclear reactors to generate electric power (figure 4.7). On March 28, 1979, valve failures caused the temperature and pressure in one reactor to rise, opening a relief valve at the top. Seconds later, all of the control rods descended, as designed to do in an emergency, to bring fission to a halt. Falling pressure in the reactor should have closed the relief valve but didn't, though, and radioactive water and steam escaped and entered the concrete containment building. Confusion and poorly designed monitoring equipment then misled the operators to cut off the flow of water into the reactor, when recooled water should have entered to reduce the heat. The fuel rods underwent partial meltdown as a result, raising the concentrations of radioactivity in the escaping water and steam, some of which entered the outside environment. The news media soon learned of the crisis, and amid uncertainty public officials came close to evacuating a wide area. After federal officials brought the reactor under control, construction of new nuclear power plants in the United States came to a halt.[26]

No one died as a result of the accident at Three Mile Island. However, an investigation by a presidential commission concluded that the failure resulted primarily from the conduct of operators, managers, and regulators, who did not employ or previously enforce standards of safe operation. Deficient training, unclear operating procedures, and a poorly designed control room contributed to the breakdown. The most important recommendation urged by the commission was better training of the people working in nuclear plants. In December 1979, the nuclear utilities created an Institute of Nuclear Power Operations in Atlanta, Georgia, a national program to certify nuclear plant operators. The Nuclear Regulatory Commission also began to enforce more stringent standards.[27]

American utilities did not build any new nuclear generating stations after Three Mile Island, and a number of stations under construction or nearing completion had to close. Some plants that met the new standards reached completion and some were able to add reactors. But the costs of new nuclear power plants had begun to rise even before Three Mile Island, and stronger regulation afterward forced expensive changes in designs and lengthened time to completion.[28] Finished plants operated safely, though, and now supply about 20 percent of America's electricity by using more of their capacity. Before 1970, nuclear plants turned about half of their rated power into

Figure 4.7
The Three Mile Island nuclear power plant near Middletown, Pennsylvania. The two tall concrete structures in the background are water cooling towers. The two round-topped concrete buildings in the foreground house reactors. The damaged reactor, TMI-2, was behind the one in front, TMI-1. Courtesy of U.S. National Archives.

actual power; since then, utilities have maintained their share of the nation's electric power supply by increasing capacity usage to between 80 and 90 percent.[29]

The Quest for Fusion

During the 1950s, fission was the most practical way to generate electric power from nuclear energy, but it was not the only way that attracted interest. The release of nuclear energy can also result from fusion, from the merging of atomic nuclei with each other. The heaviest natural element, uranium, fissions most easily, but fusion occurs most readily with the lightest element, hydrogen. Most hydrogen atoms consist of an electron and a proton. Scientists determined that an isotope of hydrogen called deuterium, with a nucleus of one proton and one neutron, could fuse best with another hydrogen isotope, tritium, having a nucleus of one proton and two neutrons, releasing about four times as much energy as the fission of a uranium-235 atom.

Fusion had several attractive features besides releasing more energy. Fission reactors needed a scarce element, uranium, while deuterium could be obtained from seawater. Tritium was harder to find, but the fusion of one deuterium and one tritium atom produced an atom of helium with a neutron left over. If the fusion reactor had lithium in it, the lithium could absorb the ejected neutron and produce more tritium, regenerating the supply. Spent fuel from nuclear fission would be radioactive for thousands of years, while the reactors and shielding in a fusion plant would be radioactive only for 100 years after ceasing operation. The only waste product from fusion was helium, which was harmless. If it could be made to work economically, fusion could replace both fission and fossil fuel as a way to generate electricity.

The challenge was to engineer a system in which a sustainable fusion reaction could take place. The enormous gravity of the Sun naturally fuses its atomic nuclei together, releasing energy. On Earth, hydrogen bombs in the 1950s used fission explosives to fuse the nuclei of small masses of deuterium and tritium in uncontrolled reactions that released more energy than fission bombs. But to sustain fusion in a controlled way on Earth would require heating deuterium and tritium to a temperature of almost 100 million degrees Celsius, six times as hot as the core of the Sun. Once it reached this temperature, the deuterium and tritium fuel had to fuse long enough to "ignite" and then burn by itself until the amount of fuel needed to be replaced.

Scientists took charge of fusion research and pursued two approaches. Since no known material could withstand the high temperature required, Lyman Spitzer, an astrophysicist at Princeton University, proposed in the 1950s to contain the reaction inside a strong magnetic field. He launched a program in what is now the Princeton

Plasma Physics Laboratory in Princeton, New Jersey, that became the center for magnetic fusion research in the United States. After experimenting with other forms of magnetic containment, scientists settled in the 1970s on a Russian design called a *tokamak* that resembled a hollow donut. Magnets surrounding the tokamak isolated and propelled deuterium–tritium fuel on a circular path inside, which an electric current then heated. The heat caused the atoms to break down into a state called *plasma* and then burn. The other approach, employed by scientists at the Lawrence Livermore Laboratory in Berkeley, California, employed lasers to compress and heat a deuterium–tritium fuel pellet. The U.S. government funded both magnetic and laser fusion research.[30]

Scientists were able to achieve short moments of fusion. Unfortunately, instabilities in the magnetically confined plasma or on the surface of the pellet in laser fusion prevented either kind of fusion from sustaining itself. The fusion also required more energy to create than the fusion energy produced in return. To overcome these problems, researchers built ever-larger machines, which required ever-increasing amounts of money. However, sustained fusion remained elusive. By the 1990s, the U.S. government decided not to increase its level of support. Magnetic fusion research is now an international effort centered in Europe, and laser fusion is now funded in the United States mainly as a way to test the fuel in nuclear warheads. Although advances in fusion research continue, past experience suggests that practical fusion is a long-term prospect.[31]

The Lessons of Nuclear Power

Rickover's nuclear fission program succeeded for three reasons. First, the enormous engineering difficulties were possible to solve in a matter of a few years, and he solved them. Second, his program met a unique need that no other source of power could meet. Only with nuclear energy could submarines have stayed underwater for weeks and months. In contrast, civilian nuclear power faced competition at the beginning from fossil fuels. Finally, in creating a form of nuclear power himself, Rickover was able to define how it should work and to set demanding standards for safe and efficient operation that he and his organization were able to enforce within the navy.

Admiral Rickover emphasized above all tying every aspect of safety to a responsible person. This rule should have applied to all dangerous energy technology, not just nuclear power. For example, in the spring and summer of 2010, a deep-sea drilling accident spilled millions of gallons of petroleum into the Gulf of Mexico. Responsibility for managing the disaster on the drilling platform Deepwater Horizon was unclear, contributing to the loss, and an investigation afterward found that design of the drilling operation and inspection before the disaster were also deficient.[32]

In recent years, America has tapped new sources of fossil energy. Vertical drilling has been the traditional method of recovering fossil fuels underground. Large deposits were not economical to recover, though, if they were mixed with other minerals. Through a process called hydraulic fracturing ("fracking"), engineers have found that they can release natural gas and oil from such mixtures by drilling vertically and then horizontally into the deposits and injecting water containing certain chemicals under pressure. Fracking has released new quantities of fossil fuel but has given rise to concern over the proper disposal of wastewater and the seismic impact of the practice in areas with sensitive geology.[33]

Nuclear energy was controversial from its beginning and remains so today. Since 1945, nuclear weapons have been credited both with preventing great power conflict and with threatening to annihilate civilization. Until the 1990s, only five countries had nuclear weapons; since then, other countries have acquired or pursued them by building nuclear reactors and facilities to enrich uranium.[34] Civilian nuclear power has been attacked and defended on economic and environmental grounds, with arguments that have changed little since the 1970s.[35] If renewable energy becomes a practical alternative to fossil fuels and to energy from nuclear fission and fusion, the latter three may be retired from service or further research.

In his development of nuclear power, Rickover continued a tradition of radical engineering innovation coming from the public sector. After 1945, aerospace engineers with public funding realized two new radical ideas: the jet engine for flight at higher altitudes and speeds, and the rocket for travel into outer space.

5   The Jet Engine and the Rocket: Whittle, Goddard, and NASA

Before World War II, experts believed that a jet engine to power an airplane would never be practical. A British air force officer, Frank Whittle, believed otherwise and proposed the idea in 1929, only to be ignored. But in 1936, with new foreign threats on the horizon, the British Royal Air Force gave Whittle some support to develop his idea. By 1939 he had a working jet engine, and two years later he had a jet airplane. During World War II, German engineers also produced jet airplanes, but jet aircraft did not play a significant role on either side during the war. Afterward, however, they quickly replaced propeller engines on combat aircraft. During the 1950s, passenger airplanes began to use jet engines for civilian air travel, and the result was a dramatic shift of passengers from rail and sea to air for travel over long distances.

The principle of rocket propulsion went back centuries earlier to ancient China, where canisters filled with explosive powder lifted rockets up hundreds of feet. Modern rocketry began when engineers started to understand how to reach much higher altitudes. The American physicist Robert Goddard privately conducted the first modern rocket launch in 1926, using a liquid fuel, after pioneering some insights that would prove crucial to later space travel. Relying in part on Goddard's ideas, German engineers developed rockets to deliver explosives during World War II. Afterward, the United States and the Soviet Union embarked on massive programs to develop rockets and piloted spacecraft. The resulting space race led in 1969 to America's historic landing of astronauts on the Moon.

The Limits of Propeller Aviation

The first powered airplanes employed piston engines to turn propellers. The combustion of gasoline in the engine moved pistons that turned a crankshaft, which rotated a shaft to which the propellers were attached. The Wright brothers used such an engine in 1903 to fly the first heavier-than-air craft in steady level flight. By the 1920s, engineers had made important improvements in aircraft design, and airplanes with propeller engines mounted in front of a fuselage could carry a pilot and a small payload. By the mid-1930s, newer airplanes had all-metal frames and closed streamlined bodies, with monoplane instead of biplane wings, and some airplanes had an engine mounted on each wing. The innovation of new refining processes in the 1930s improved gasoline, making it safer and more powerful as an aviation fuel.[1]

By the 1930s, though, engineers realized that propeller aviation had limits. Propeller tips rotated at speeds faster than the forward speeds of the airplanes they propelled, and experiments showed that as propeller speeds approached the speed of sound, the resistance of the air on their tips rose sharply. As a result, the maximum speed that propeller-driven aircraft could reach was 350–450 miles per hour (563–643 km/hr).[2] Propeller planes also had a ceiling to how high they could fly. The cruising altitude of the Douglas DC-3, a passenger airliner introduced in 1936, was 11,000 feet (3,352 m) at a speed of 192 miles per hour (308 km/hr), with a service ceiling of 23,000 feet (7,010 m).[3]

The jet engine joined two existing engineering ideas, the turbine and air compression, in a new kind of engine that dispensed with pistons. During the 1800s, turbines delivered water to power machines in textile factories, and later turbines served as electric generators attached to dams. Charles Parsons in England invented a stationary turbine in 1884 that used steam instead of water to turn a bladed shaft. The Parsons steam turbine was more efficient than any piston steam engine in existence. However, steam turbines were too heavy to power motor vehicles or airplanes.[4] In 1907, an American engineer with General Electric, Sanford Moss, designed a stationary turbine that burned a liquid fuel to drive a shaft. But his machine required four times as much fuel as a piston engine to deliver the same amount of horsepower per hour.[5]

Engineers in the early twentieth century were able to make use of air compression in airplane engines. During World War I, aircraft builders designed piston engines with intakes designed to constrict the flow of air into the engine, compressing the air. This compression, called *supercharging*, made the air flow more evenly when it mixed with fuel, improving combustion, and the added oxygen raised the altitude at which airplanes could burn fuel and fly. However, piston-engine combat airplanes could not sustain a cruising altitude much above 40,000 feet (12,192 m).[6]

In 1926, Alan A. Griffith, a researcher at the Royal Aircraft Establishment, a government laboratory in Britain, proposed using compressed air and a turbine for propulsion. His idea was to compress air before mixing it with a liquid fuel and then to have the exhaust gas from combustion rotate a bladed shaft (a turbine) that would rotate propellers attached in front. Griffith's idea still relied on propellers, though.[7] The Langley Aeronautical Laboratory in the United States, the American counterpart to the Royal Aircraft Establishment, chose to focus in the 1920s and 1930s on improving propeller aviation with piston engines, as did the private aircraft industry worldwide.

Expert opinion also discouraged the pursuit of a true jet engine, in which compressed air would combine with a liquid fuel, ignite, and exhaust the resulting gases through a nozzle, moving an airplane forward in reaction to the thrust of the exhaust jet. In 1924, an American physicist with the U.S. National Bureau of Standards, Edgar Buckingham, wrote a report concluding that an engine of this kind would require four times as much fuel to achieve the same thrust at 250 miles per hour (402 km/hr) as a conventional propeller engine. In making his calculations, though, he included the weight of an onboard piston engine to compress the air, and he did not consider the higher speeds and altitudes at which a jet engine might have been more efficient.[8] Expert judgment in the United States remained skeptical of jet propulsion as late as 1940. In that year, a committee of experts appointed by the U.S. National Academy of Sciences concluded that an airplane powered by a jet engine was not a practical prospect.[9]

Frank Whittle and the Jet Insight

A true jet engine was the innovation of Frank Whittle (1907–1996), who designed and built such an engine in Britain in the late 1930s. Whittle grew up in the town of Leamington, England, where his father was a mechanic. In his youth, Frank Whittle read widely about aviation and mastered the working of an internal combustion engine owned by his father. At age 16, after twice failing local physical examinations to enlist in the Royal Air Force (RAF), he put himself through physical conditioning and applied for an apprenticeship at RAF Cranwell, an air base in the next county. He qualified and trained to be an airplane mechanic. Three years later, having excelled in his work, he entered the RAF cadet college, also at Cranwell, for training to be a commissioned officer (figure 5.1). Most of the other cadets had attended boarding schools instead of coming from the ranks of apprentices, but Whittle showed exceptional ability in his academic work and he proved an excellent pilot. Daredevil flying was held to be his only defect and he graduated in 1928 at the age of 21.[10]

Whittle wrote a thesis on how to fly at higher altitudes, concluding that it should be possible if propeller engines could attain greater speeds with the help of a gasoline-fueled turbine in back.[11] A year after graduation, though, it occurred to Whittle to dispense with the propeller altogether and just rely for propulsion on the exhaust jet from a combustion chamber that burned a liquid fuel and a compressed flow of air.

Figure 5.1
Frank Whittle (*back row, center*) as an officer cadet. Courtesy of Archive, Institution of Mechanical Engineers, United Kingdom.

"Once the idea had taken shape," he noted later, "it seemed rather odd that I had taken so long to arrive at a concept which had become very obvious and of extraordinary simplicity."[12] Whittle believed that he could design a jet engine efficient at an altitude of 70,000 feet (22,300 m) with a speed of 500 miles per hour (804 km/hr). The cold temperatures at this height would give more power with a given amount of energy than at lower altitudes. He filed a patent claim in 1930 outlining a design for such an engine.[13]

The British Air Ministry, the government department in charge of the Royal Air Force, turned down the proposal, as did the expert Alan Griffith, who agreed to review it and found an error in one of Whittle's main calculations (later found not to be wrong). Several leading airplane manufacturers also rejected the idea, reluctant to take new risks in the deepening depression of the early 1930s. In 1934, Whittle obtained admission to Cambridge University, where he completed a three-year honors program in mathematics and science within two years. But in January 1935, lacking the money to renew his patent and discouraged about its prospects, he allowed the patent to lapse.[14]

A few months later, however, two Royal Air Force officers who had returned to civilian life persuaded a small investment bank to provide £10,000 ($50,000 in 1936, or more than $800,000 in 2010) to support Whittle in the development of his jet engine. Whittle and his partners incorporated as Power Jets, Ltd., with Whittle's shares held by the Royal Air Force. With the possibility of a second world war on the horizon, his superiors in the RAF decided to see what Whittle could do and they exempted him from regular duty. Whittle made some design changes to his lapsed patent design, so that his partners could obtain a new patent. The British Thomson-Houston company, which made engines, agreed to manufacture the equipment.[15]

Engineering the Jet Engine

Frank Whittle faced an extraordinary set of challenges even with his newfound support. In the prototype that he now designed, air would enter an opening in the front of the engine and would hit a bladed wheel mounted on a shaft (box 5.1). This wheel, the compressor, would deflect and compress the air into a funnel that led to a combustion chamber, where the air and a liquid fuel would ignite in a continuous burn. The hot gases would then pass through another wheel with blades, the turbine, mounted at the rear of the shaft. The gases going through the turbine wheel would rotate the shaft and turn the compressor blades in front. At the same time, the gases passing through the turbine would leave the engine as an exhaust jet and give the airplane its forward thrust.

Box 5.1

The Whittle Turbojet Engine

Engine shaft with larger compressor wheel (*center*) and smaller turbine wheel (*right*).

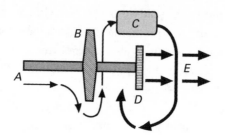

Whittle engine diagram (simplified).

Frank Whittle's first turbojet engine consisted of three basic components (*above, top*): a shaft, a compressor wheel, and a turbine wheel (engine casing omitted).

The diagram (*above*) outlines how the engine worked. Air would flow into the engine at the air intake opening *A*. The compressor wheel *B* inside the intake would compress and divert the airflow into the combustion chamber *C*, where the air and an injected fuel would ignite. The resulting hot gases would then loop through piping around the shaft to the turbine wheel *D*. The gases went through the angled blades of the turbine wheel, rotating it and the attached shaft holding the compressor wheel. The exhausted gases at *E* would provide thrust to propel the airplane forward, and the rotating compressor wheel would draw more air into the engine.

By diverting air around the compressor wheel, Whittle's engine was said to employ a kind of air intake called *centrifugal flow*.[a] Later jet engines used *axial flow*, in which air went through the blades of several compressor wheels, ignited in a single chamber behind the wheels, and then exhausted the hot gases through the turbine.[b]

The jet engine eliminated the need for propellers at the front of an airplane. However, engineering a design so that all of the components would work together proved extraordinarily difficult at first, and Whittle tried many different versions before finally settling on a design and getting it to work acceptably.

Source: Image of shaft and wheels from Air Commodore F. Whittle, "The Early History of the Whittle Jet Propulsion Gas Turbine," James Clayton Memorial Lecture (London: Institution of Mechanical Engineers, 1945). Courtesy of Archive, Institution of Mechanical Engineers, United Kingdom. *Notes:* (a) Whittle designed his engine with centrifugal flow because he was familiar with earlier compressors that used the method. (b) For a diagram of an axial-flow engine, see figure 5.6 below.

Figure 5.2
Whittle's first experimental jet engine (1937). Courtesy of Archive, Institution of Mechanical Engineers, United Kingdom.

To reach 500 miles per hour with an aircraft weight of about 2,000 pounds (907 kilograms, or kg) would require a shaft speed in the engine of about 17,000 revolutions per minute (rpm). This was not unusual in a large stationary turbine but had never been attempted in a flying machine; the propellers on a Douglas DC-3 rotated at 2,000 rpm. The stresses on moving parts together with the much higher combustion temperatures would have no precedent in any previous aircraft engine. Owing to limited funds, Whittle could not build and test his components separately, which meant that he had to assemble and test the engine as a single machine (figure 5.2).[16]

In April 1937, the engine was ready on a test bench in a large room at the British Thomson-Houston factory, with the exhaust nozzle pointed out of a window. After bringing it up to 2,000 rpm with an electric motor, Whittle switched to a diesel oil supply and listened as the shaft speed accelerated with a terrifying scream. Everyone on the floor fled the building except Whittle, who tried to shut off the engine only to have it continue to accelerate. Finally, after reaching half of its intended speed, it began to slow down. Afterward, Whittle found that a fuel leak had fed the acceleration until the leaked fuel had been used up. This was the first of a nearly unending series of frightening mishaps, as Whittle and a handful of engineers and mechanics tried to fix one problem after another over the next two years. Testing moved to a separate building

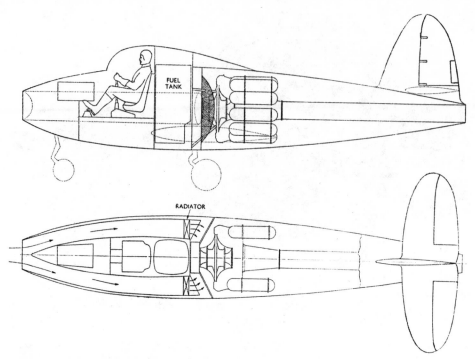

Figure 5.3
Gloster E.28/39 airplane design. Upper drawing is a side view that shows the redesigned engine using multiple combustion chambers. Lower drawing is a top-down view that shows air intakes at the front. Courtesy of Archive, Institution of Mechanical Engineers, United Kingdom.

for safety and resumed, after some additional funding, on a modified version of the engine, with 10 smaller combustion chambers instead of a single big one. Injection of the fuel still caused difficulties. But finally, on June 30, 1939, the new engine reached 16,500 rpm in a successful bench test witnessed by officials from London.[17]

Britain entered World War II on September 3, and the Royal Air Force now gave Whittle the backing to develop an experimental jet airplane. A small firm, the Gloster Aircraft Company, worked with him to design what the company designated the E.28/39 airplane. The design placed a jet engine behind the pilot, in the rear of the fuselage, with an inlet on each side of the plane in front of the pilot that fed air into the engine. A single nozzle at the end of the fuselage let out the jet exhaust (figure 5.3).

Difficulties continued, chiefly the need for a more durable metal in the turbine section, until a specialty firm invented a nickel alloy, Nimonic-80, that could withstand the harsh temperatures and gases of the exhaust flame. In May 1941, the jet plane was finally ready. Whittle got into the cockpit at Cranwell, turned on the ignition himself, and then got out and gave the plane to a test pilot, who conducted a successful flight test at 17,000 rpm. The test demonstrated the superior speed and maneuverability of the plane compared with the best British propeller warplanes (figures 5.4 and 5.5).[18]

The success of Whittle's flight test led the Royal Air Force to order a jet plane for use in combat, and the Gloster Meteor, with a jet engine on each wing, entered service in 1944. However, the plane was deployed mainly to intercept jet-propelled German cruise missiles called V-1 bombs. In 1941, the British shared the Whittle engine design with the United States. Engineers at General Electric quickly made a jet engine that the Bell Aircraft Company built into an airplane, the Airacomet, which flew in October 1942. Lockheed Aircraft also built a jet plane, the Lockheed P-80. However, senior U.S. officers were hesitant and the new planes did not serve in combat during the war.[19]

Germany also developed jet airplanes. A German engineer, Hans Joachim Pabst von Ohain, conceived the idea of a jet engine in 1932 and secured the backing of Ernst Heinkel, the founder of a company that would be a major supplier of warplanes to the German air force in World War II. With this support, Ohain was able to build and fly an experimental jet airplane by August 1939. Other German aircraft firms soon designed jet planes of their own. The Nazi leadership delayed jet aircraft production until late 1943, though, and then used the planes as low-level fighter-bombers rather than as high-altitude interceptors for which they were better suited. By the time they were used as interceptors in late 1944, it was too late for them to affect the outcome of the war.[20]

From Turbojets to Turbofans

After 1945, the United States, its major allies, and its Cold War adversary, the Soviet Union, converted their combat air forces to jet propulsion. Postwar military jet airplanes introduced two improvements. First, in the engine itself, a different kind of air compression, known as axial flow, proved more efficient than the centrifugal-flow compression developed by Whittle. In axial flow, air went through multiple compressor wheels attached to the shaft instead of deflecting around a single compressor wheel. Blades on each of these wheels formed a row that alternated with a row of blades attached to the engine casing that did not rotate (figure 5.6). The arrangement stabilized and compressed the air before it entered the combustion chamber. Axial flow enabled engines to be narrower and gave a simpler path for the air to go through.[21]

Second, the wingspan of postwar jet planes changed to a swept-back inverted "V" shape, starting in America with the F-86 Sabre jet that began service in 1949 (box 5.2) (figure 5.7). The swept-back wings gave the airplane less air resistance as the plane

Figure 5.4
Group Captain Frank Whittle during World War II. Courtesy and Crown Copyright: Imperial War
Museums, London.

Figure 5.5
Gloster E.28/39 experimental jet airplane (with a centrifugal-flow engine inside). Courtesy and Crown Copyright: Imperial War Museums, London.

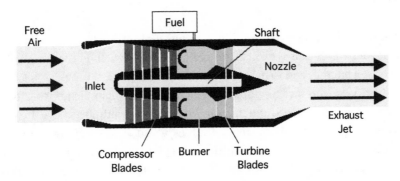

Figure 5.6
Diagram of an axial-flow engine, which replaced the centrifugal-flow engine after World War II. Courtesy of NASA Glenn Research Center and Tom Benson (adapted from original drawing).

Box 5.2
The Swept-Back Wing

After 1945, as the speed of jet-propelled aircraft approached and then exceeded the speed of sound, the design of jet airplane wings changed.

At these higher speeds, airplanes in which the wings were perpendicular to the fuselage encountered much higher air resistance or drag. Above the speed of sound, this drag produced a shock wave, a sudden increase in compressibility of the air. The photo below shows the shock wave produced by the perpendicular wingspan of a model airplane in a wind tunnel against a wind 1.72 times the speed of sound:

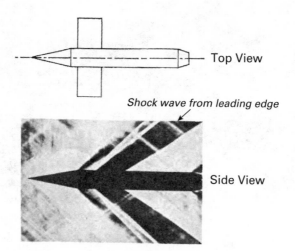

Top View

Shock wave from leading edge

Side View

Perpendicular airplane wings. Shock wave over and under wings is severe.

approached and then exceeded the speed of sound. The success of military jet airplanes with swept-back wings caused civilian airlines to begin to switch in the late 1950s from propellers to jet engines and to design airliners with swept-back wings.[22]

The British were the first to employ jet engines on civilian aircraft to offer long-distance passengers faster speed and the comfort of flying in altitudes above the weather. During the war, Geoffrey de Havilland, a leading British aircraft manufacturer, made a jet engine based on the Whittle design, and in 1952, the de Havilland firm introduced the Comet, a passenger airplane with two jet engines built into each of its wings. Over the next two years, though, four Comet airplanes crashed as a result of failures in the airframe. Four years later, de Havilland introduced a redesigned and safer version of the

Box 5.2 (continued)

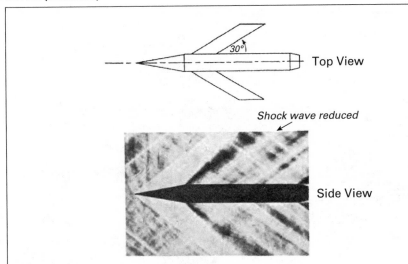

Swept-back airplane wings. Drag is much less severe.

By designing the wings with a swept-back arrangement, as shown in the photo above, the drag of the airplane and the resulting shock wave could be greatly reduced at the same supersonic speed. The change left only a much smaller drag caused by the frontal area of the model.

The ratio of an aircraft's speed to the speed of sound is known as its Mach number. The models above faced a wind with a Mach number of 1.72.

Source: Theodore von Kármán, *Where We Stand: A Report of the AAF Scientific Advisory Group* (Dayton, OH: Wright Field, 1945). Images from ibid., 6.

Comet. But in 1958, the Boeing Company in the United States introduced the Boeing 707, an airliner with four jet engines that could carry up to 189 passengers on intercontinental flights, twice as many passengers as the Comet (figure 5.8).[23]

The Boeing plane, followed by the Douglas DC-8, captured most of the market for civil jet airliners. The result was a revolution in transport on the ground as well as in the air. By 1980, rail travel in the United States had dropped by two-thirds from its level in 1950, from 32 to 12 billion passenger-miles annually, while air travel increased from 10 to 219 billion passenger-miles. A shift to private automobiles accounted for most of the decline in rail travel, but the rise in air travel reflected the preference of passengers for the speed of jet airliners to that of long-distance trains.[24]

Figure 5.7
A U.S. Air Force F-86 Sabre jet (*left*), pursuing a later U.S. Navy jet plane. Courtesy of NASA Armstrong Flight Research Center.

The dramatic growth of civilian jet air travel led to a change in the kind of jet engines installed on airplanes. Turbojet engines provided military combat planes with the high speed that they needed for short intense engagements, but their engines consumed fuel very quickly. Passenger airliners did not need such high speed but required more thrust to carry more people. The *turbofan* engine met this civilian need (box 5.3). In a turbofan, a larger casing enclosed a turbojet engine, allowing space for cold air to flow around the engine. The shaft had large fan blades mounted on the turbine shaft before the compressor section. Some of the air pulled in by the fan blades went into the turbojet and the rest of the air remained cold and went around the engine. A turbofan engine did not reach the speed of a pure turbojet but had the added thrust of the cold air exiting in back. Using kerosene instead of gasoline, a turbojet consumed about 1 pound of fuel per pound of thrust per hour, while a turbofan needed just over half as

Figure 5.8
Boeing 707 in flight. Courtesy of Boeing Image Archives.

much kerosene fuel per pound of thrust per hour.[25] In addition to the fuel economy, turbofans made turbojet engines quieter, and for these reasons, as well as the added thrust, civilian airliners changed to turbofans. Introduced in 1970, the Boeing 747, with four turbofan engines, was able to carry 382 passengers and fly 608 miles per hour (978 km/hr) at 30,000 feet (9,144 m). The Boeing 707 flew 189 passengers at almost the same height and speed with its four turbojet engines; the greater thrust of the four turbofans enabled the 747 to carry twice as many people.[26]

Passenger jet air travel expanded the need for airports, which stimulated new construction of terminal buildings. Runways resembled paved highways, although their concrete needed to be steel-reinforced to preserve their smoothness and strength against the much heavier weight of large jet airliners. A new network of air flight paths paralleled the new network of interstate highways, but air travel had ground-based

Box 5.3

From Turbojet to Turbofan

At first, jet engines after 1945 were axial-flow *turbojets*, in which air entered an intake passage. Compressor wheels compressed the air before feeding it into a combustion chamber behind them, where the air mixed with fuel, ignited, and produced an exhaust jet (*below*).

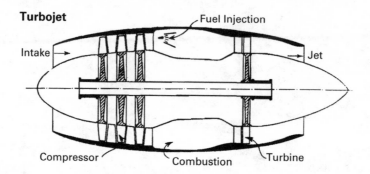

Military aircraft relied on turbojet engines to achieve high speed for short periods of time in the air. In contrast, passenger airliners cruised at lower speeds but needed to stay aloft for many hours at a time. The turbofan engine met this civilian need.

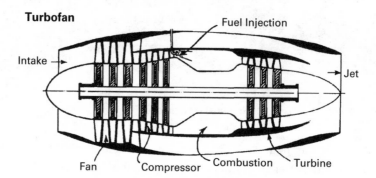

In a *turbofan* engine (*above*), a turbojet engine stood inside a larger casing. Fan blades mounted inside the turbofan inlet took in air. Some of this air went into the turbojet engine, where it compressed, mixed with fuel, and produced combustion and an exhaust jet. The rest of the air went around the engine and out the back. This cold air did not add speed but augmented the engine's thrust and saved fuel.

Passenger airliners converted to turbofan engines after 1970 to carry more people and other weight. The new engines were also quieter than turbojets. Military airplanes eventually obtained turbofan engines that could achieve high speed as well as high thrust.

Source: Theodore von Kármán, *Where We Stand: A Report of the AAF Scientific Advisory Group* (Dayton, OH: Wright Field, 1945). Images from ibid., 19–20.

radio control to ensure safety and efficiency.[27] The rise of jet air travel aided the shift of American civilization from central cities to suburban areas, which had room for airports.

Frank Whittle was not the only aviator to encounter resistance to the idea of jet propulsion. In 1927, Lieutenant Sidney Parahm Vaughn, an American naval officer stationed at Pearl Harbor, Hawaii, requested funds to develop a turbojet engine. Experts at the National Advisory Committee on Aeronautics, the federal body that managed the Langley Laboratory, reviewed his proposal and turned him down, arguing that such an engine was not practical.[28]

The jet engine required for its invention a conceptual jump, from using a compressor and turbine to drive a propeller to using them without a propeller to achieve forward thrust. Once early jet aircraft had proved such a jump, other engineers embraced and improved the breakthrough. The civil jet airliner reached maturity in the 1970s and 1980s. New materials and engines may improve passenger airplanes further, but turbofan engines seem likely to remain the means of powering long-distance civil air travel for the foreseeable future.

Robert Goddard and the Modern Rocket

Modern rocketry and space travel originated in the minds of a Russian, Konstantin Tsiolkovsky (1857–1935), and an American, Robert Goddard (1882–1945). Affected by partial deafness, Tsiolkovsky was self-educated but was able to work in adult life as a secondary school teacher. The novels of Jules Verne inspired him to imagine how space travel might be possible; instead of Verne's idea of firing a spaceship from a giant cannon, Tsiolkovsky proposed in 1903 to use a rocket powered by liquid fuel. He proposed ignition of liquid oxygen and liquid hydrogen, with a tapered nozzle at the rear of the rocket to exhaust the expanding combustion gases. The explosive force of this expansion would propel the rocket. Tsiolkovsky was never able to build an experimental rocket, though, and at the time his ideas were unknown outside Russia.[29]

The pioneer of rocketry in America was Robert Goddard, who also suffered in his youth from poor health. In 1899, at age 17, Goddard climbed a tree and imagined going to the planet Mars; thereafter he made space travel his goal in life. In 1911, he earned a doctoral degree in physics from Clark University in Worcester, Massachusetts, and after study at Princeton University he joined the Clark faculty in 1914. Goddard independently arrived at conclusions similar to Tsiolkovsky's. However, Goddard also began to test model rockets and he established principles that would prove essential to space travel.[30]

An airplane flies by balancing four forces: it must generate sufficient lift under its wings to carry its weight, and it must develop sufficient thrust to overcome the resistance of the air, also known as aerodynamic drag. A rocket is different. Instead of flying as an airplane would, at a constant altitude through air of constant density, a rocket bound for outer space moves vertically to altitudes where air density quickly lessens and disappears if the rocket leaves the atmosphere. Drag is not a significant factor, and the principal task of the rocket is to develop sufficient thrust to lift its weight to the desired altitude.

Goddard grasped these ideas. He also recognized that a rocket with a single fuel compartment would carry excess weight as the rocket exhausted its fuel. Although the idea originated earlier, Goddard developed in practical form a rocket with multiple stages to sustain flight at higher altitudes. He envisioned a series of rockets mounted on top of each other. When each rocket in the stack had expended its fuel, it could be ejected to release weight. Goddard's most important insight was that a rocket could move in a vacuum. Sir Isaac Newton's third law of motion stated that for every action there was an equal and opposite reaction. However, it was believed by many at the time that the exhaust gases from a rocket engine needed to react against a medium of some kind, such as air, to move the rocket forward. Goddard showed experimentally that rocket propulsion was possible in a vacuum. His published reports in 1919 attracted wide attention, although his confidence about space travel also invited some public skepticism.[31]

Goddard began experiments with small rockets filled with gunpowder. He soon realized that a mixture of gasoline and liquid oxygen would be more efficient, and with this fuel mixture, he launched a small rocket from his Aunt Effie's nearby farm on March 16, 1926 (figure 5.9). The rocket rose 41 feet in 2.5 seconds and landed in a cabbage patch. The noise from a larger rocket launch on the farm three years later frightened neighbors into calling the police, though, and Goddard was ordered not to launch any more rockets in Massachusetts. The Smithsonian Institution had backed his research in the early 1920s but now cut it back.[32]

Fortunately, Charles Lindbergh, whose historic solo airplane flight across the Atlantic Ocean in 1927 had electrified the world, became an admirer of Goddard and persuaded the aviator and philanthropist Harry F. Guggenheim to underwrite the rocket researcher's work at a site near Roswell, New Mexico.[33] Goddard moved to Roswell in 1930, where plentiful desert gave him the land that his work required. Here with some assistants he designed and built rockets 20 feet long, with better mechanisms to control the flow of fuel and steer the rocket in flight (figure 5.10). By 1937, he was able to launch a rocket to an altitude of about 9,000 feet (2,743 m). But with only private

Figure 5.9
Robert Goddard with his 1926 test rocket in launch frame. Courtesy of NASA Marshall Space Flight Center.

Figure 5.10
Robert Goddard in his workshop, Roswell, New Mexico, 1935. Courtesy of Clark University Archives.

funding Goddard could do no more. During World War II, he advised the U.S. Navy on rockets, but he died in 1945 before the postwar age of space exploration began.[34]

Goddard was not the only American to take a serious interest in rocketry in the 1930s. A small circle of graduate students at the California Institute of Technology (Caltech) began to design rocket engines and received Guggenheim support in 1939. Under the direction of Theodore von Kármán, a leading aerodynamicist on the Caltech faculty, the students worked out their ideas in theory, but early practical tests were not well executed. During the war, however, the group used rocket engines to assist the takeoff of propeller airplanes, and after the war the Caltech group built a working rocket and helped found the Jet Propulsion Laboratory in Pasadena, California.[35]

A more substantial rocket program developed in Germany. In a 1923 book, Hermann Oberth (1894–1989) embraced the idea of a multistage rocket and helped stimulate amateur interest in rocketry.[36] After Hitler's coming to power in 1933, civilian rocket research became illegal in Germany but the German army funded a rocket program for military use. The key figure in this program was Wernher von Braun (1912–1977), a recently graduated aeronautical engineer, who studied Goddard's work and supervised the design and testing of rockets at a site on the Baltic German coastline, Peenemünde. During World War II, the British bombed the Baltic base, causing the Germans to move rocket assembly to an underground site in central Germany, where slave laborers from a nearby concentration camp assembled the rockets. Von Braun tolerated the maltreatment of these laborers.[37]

In September 1944, Germany began launches against England of a single-stage rocket, the V-2, fueled by alcohol and liquid oxygen. (The V-2 designation followed launches of a jet-propelled cruise missile, the V-1, in June.) The V-2 rocket traveled at supersonic speed to an altitude of 55 miles (88 km) and was able to carry an explosive payload to targets 200 miles (320 km) distant. German rockets killed British civilians but had no impact on the military course of the war. In 1945, von Braun and most of his staff were able to surrender to advancing American troops, who moved them and a supply of surviving V-2 rockets to the United States.[38]

NASA and the Moon Landing

After World War II, the U.S. armed forces began to develop rockets, as did those of their Cold War adversary, the Soviet Union. Colonel Chuck Yeager of the U.S. Army, flying a rocket-powered airplane, the Bell XS-1, broke the speed of sound in 1947.[39] American efforts lacked urgency, though, until October 4, 1957, when the Soviet Union launched a small satellite, Sputnik-1, into Earth orbit. Fear of falling behind in space prompted President Eisenhower and Congress to increase federal funding for aerospace defense and higher education. A federal civilian body, the National Advisory Committee on Aeronautics (NACA), merged with the Army Ballistic Missile Agency to form a new civilian agency, the National Aeronautics and Space Administration (NASA), with responsibility for space rockets to carry astronauts. The U.S. Air Force separately began to develop missiles of intercontinental range to deliver nuclear warheads, while the U.S. Navy developed Polaris missiles to do the same from nuclear-powered submarines.[40]

On May 5, 1961, America launched astronaut Alan Shepard into outer space. However, on April 12 the Russians had already launched the first human into space, Yuri Gagarin, who also orbited the Earth.[41] The new U.S. President elected in 1960, John F. Kennedy, responded by setting a dramatic goal for the United States. On May 25, 1961, in a message to Congress, Kennedy declared: "I believe that this nation should commit itself to achieving the goal, before this decade is out, of landing a man on the moon and returning him safely to the earth."[42] Kennedy charged NASA with carrying out this historic mission.

A lunar flight had two basic engineering requirements. One was a guidance system capable of extraordinary precision: a small error in navigation could cause an outbound spaceship to crash into the Moon, or on its return journey to miss the Earth. Fortunately, a mechanism developed in the 1950s by the engineer Charles Stark Draper at the Massachusetts Institute of Technology (MIT), called inertial guidance, could detect when an airplane deviated from a level flying path and could do so without reference to the ground. The device could work on spacecraft and could supplement ground control using radar. Coupling this device to a compact computer would make navigation in outer space possible. In the early and middle 1960s, with the help of newly invented microchips (see chapter 7), Draper's Instrumentation Laboratory at MIT designed a compact onboard computer to navigate spaceships.[43]

The other requirement was a rocket powerful enough to lift a payload of 100 U.S. tons (90.7 metric tons) beyond the Earth's gravity. NASA's Marshall Space Flight Center near Huntsville, Alabama, took over the design of such a rocket under the direction of Wernher von Braun.[44] The engineers at the Marshall Center concluded that a single rocket able to leave the Earth, land on the Moon's surface, and then return to Earth would be too large to be practical. In its place, von Braun and his team recommended an approach, called Earth Orbit Rendezvous, in which several rockets would lift components of a spaceship into orbit for assembly. From here, the spaceship could be sent to the Moon and back. However, an engineer at the Langley Laboratory outside Washington, John Houbolt, argued instead for a third approach called Lunar Orbit Rendezvous (figure 5.11). Under this approach, a rocket would boost onto a lunar journey a

Figure 5.11
John Houbolt explaining Lunar Orbit Rendezvous. Courtesy of NASA Headquarters.

Command Module, an attached Service Module to supply it, and a separate lander, the Lunar Excursion Module. Once in lunar orbit, two of the three astronauts on the mission would enter and detach the Lunar Module, land on the lunar surface, and then return to the orbiting command ship, which would carry them home, leaving the other two modules in space. Houbolt's approach was more economical than Earth Orbit Rendezvous, and NASA (and von Braun) accepted it in 1962.[45] To work, though, astronauts would have to bring three interconnected modules and would need to learn how to perform docking maneuvers in the vacuum of space.

NASA built a series of new launch sites at Cape Canaveral, Florida, where prevailing winds and vast ocean expanses to the east made early rockets practical to launch and splash down.[46] Project Mercury, which began in 1959, launched astronauts individually into space from 1961, and one of them, John Glenn, orbited the Earth a year later.

Figure 5.12
The Saturn V rocket first stage, showing its five engines. Courtesy of NASA Marshall Space Flight Center.

Project Gemini followed in 1965–1966, in which pairs of astronauts carried out operations in Earth orbit, including extra-vehicular activity and docking. Work on Project Apollo, the program for a Moon landing, began in 1962. Even with the more economical plan of Lunar Orbit Rendezvous, von Braun and his engineers would need to design and build a three-stage booster rocket of unprecedented size and power.[47]

America's largest missile in the 1950s was the Jupiter rocket, and engineers at the Marshall Space Flight Center named the new larger rocket for Apollo missions after the next planet, Saturn. The four turbofan engines of a Boeing 747 delivered a total thrust averaging 180,000 pounds (81,646 kg). The first stage of the Saturn rocket would need to have five F-1 engines, each capable of delivering about 1.5 million pounds of thrust (680,388 kg) (figure 5.12). NASA relied on private contractors to build each stage: Boeing would build the first stage, North American Aviation the second, and Douglas Aircraft the third. Rocketdyne would build the five engines of the first stage, the five smaller J-2 engines of the second, and the single J-2 engine that would power the third.[48]

In the first stage, two separate tanks would hold liquid oxygen and a highly refined form of kerosene, which would combine in each engine combustion chamber and ignite, releasing a hot blast of gases out of each nozzle below. To enter the combustion chamber, fuel went through a circular "injector" disk with holes in it, and tests on an H-2 engine, a smaller prototype, revealed that combustion would be difficult to control. By enlarging the holes, rearranging them, and placing baffles (ridges) on the side of the disk where combustion would occur, the engineers were able to guide the fuel into the combustion chamber to produce a smooth flow. When the engineers scaled up the H-2 to create the F-1 engine, however, the redesigned injector disk didn't work. Failure to solve the problem soon threatened the entire lunar mission. With the help of other NASA centers and several universities, and through trial and error, Rocketdyne finally found a design of the disk and other parts that produced acceptable combustion. The engineers and scientists involved never fully understood what went on in the combustion chamber. Modern engineering often relies on such "black box" solutions, in which what happens during a process doesn't need to be fully understood as long as engineers know reliably what goes into the process and what comes out of it.[49]

The second stage raised another problem. NASA decided that the engines of the second and third stages would combine liquid oxygen with liquid hydrogen, because the latter could deliver much more energy per pound than kerosene, saving weight. Liquid oxygen was possible to store at a temperature of −297°F (−182.7°C), but liquid hydrogen had to be stored at a temperature of −423°F (−257.2°C). Panels of synthetic insulating foam applied to the outside surfaces of the rocket (to keep the fuel inside cold) trapped air and turned it into liquid oxygen, causing the insulation to lose adhesion. The engineers at Douglas Aircraft avoided the problem in the third stage by applying insulating foam tiles to the inside of the rocket walls. The larger second stage, however, needed the foam on the outside.

To preserve adhesion, engineers ran helium through grooves on the underside of the insulation, because helium didn't condense at the low temperature of the hydrogen. The solution was not completely effective, though, and eventually the engineers realized that if they sprayed insulation onto the outer tank surfaces, they could eliminate the trapped air altogether. Similar outcomes are not uncommon in engineering, when engineers try to solve a design flaw by first devising an adjustment to make the flawed design work better, only to find later a design or approach that eliminates the flaw.[50]

The work on the Saturn rocket required long hours of effort by engineers, technicians, and workers on a tight schedule. The rocket evolved through several changes in size and required thousands of tests for its millions of parts. NASA and the contractors stressed reliability and tested everything they could exhaustively. The final three-stage booster rocket, designated the Saturn V (Saturn Five), was 393 feet (120.3 m) high and had to be assembled standing up in the huge Vehicle Assembly Building at Cape Canaveral. An enormous tractor would move it to a nearby launch pad, where it would take off into space. While work on the rocket stages advanced, other engineers and workers completed the spaceships that would leave the Earth for the Moon: the Command Module to house the astronauts, the Service Module, and the Lunar Module.[51]

In January 1967, a test of the Command Module with astronauts inside ended in tragedy when an electrical short-circuit ignited the oxygen-rich air in the cabin and killed the crew. The tragedy led to changes to improve crew safety.[52] An unpiloted Saturn V rocket finally launched successfully in November 1967. After several more unpiloted flights, a piloted test in Earth orbit of the Command and Service Module (lifted by a smaller rocket), in October 1968, went well.[53] On December 21, 1968, a Saturn V lifted the three astronauts of the Apollo 8 mission on a journey to orbit the Moon, which they circled 10 times before returning home on December 27. The Apollo 8 astronauts took photographs of an "Earthrise" over the Moon's horizon (figure 5.13). An Apollo 9 mission tested the Lunar Module in Earth orbit and Apollo 10 went back to the Moon, where its astronauts took the Lunar Module to within 50,000 feet (15,240 m) of the lunar surface before returning to the command ship and proceeding home.[54]

On the morning of July 16, 1969, a giant tractor moved a Saturn V rocket to its launch pad (figure 5.14) and then lifted into space the three astronauts of Apollo 11, the historic mission to land on the Moon. After orbiting the Earth twice, the third-stage rocket fired again to insert the Command and Service Module onto its lunar course. The astronauts then maneuvered the Command Module to extract the Lunar Module from the third stage, jettisoned the third stage rocket, and proceeded to the Moon. After a single mid-course correction, the spaceship entered lunar orbit on July 19, and on July 20 Neil Armstrong and Edwin Aldrin began the 12-minute descent to the Moon's surface in the Lunar Module, leaving Michael Collins in lunar orbit. Seconds before landing, Armstrong observed from a window that the programmed landing site was strewn with boulders. With a few seconds of fuel left, he switched from automatic to manual control of the lander and set it down on clear terrain nearby in the Sea of Tranquillity.

Armstrong, the mission commander, was first to set foot on the Moon, as one-fifth of the world's people watched on television (figure 5.15), and Aldrin followed. The two astronauts spent the next six hours taking photographs, gathering rock and soil samples, and setting up experiments that they would leave behind. The astronauts then returned to the Lunar Module, which separated from its base and returned to

Figure 5.13
The first "Earthrise" photograph, taken by the crew of Apollo 8 (in monochrome at this angle). Courtesy of NASA.

Figure 5.14
Moving the Saturn V rocket for the Apollo 11 mission from the Vehicle Assembly Building (*left*) to the launch pad at Cape Canaveral. Courtesy of NASA.

the Command Module orbiting the Moon. The astronauts transferred their geological samples to the command ship. Three days later, on July 27, the Command Module detached from the Service Module, sending the latter toward the Sun, and reentered the Earth's atmosphere, where its heat shield deflected a fiery two-minute plunge that reached 5,000°F (2,760°C). A parachute opened to brake splashdown in the Pacific Ocean, where a U.S. Navy aircraft carrier retrieved the module and the crew.[55]

The Apollo Program in Perspective

Six more Apollo missions went to the Moon, of which five landed; an oxygen tank explosion on the way forced the crew of Apollo 13 to cut short their mission and return to Earth. But the missions after Apollo 11 were anticlimactic: public interest in space

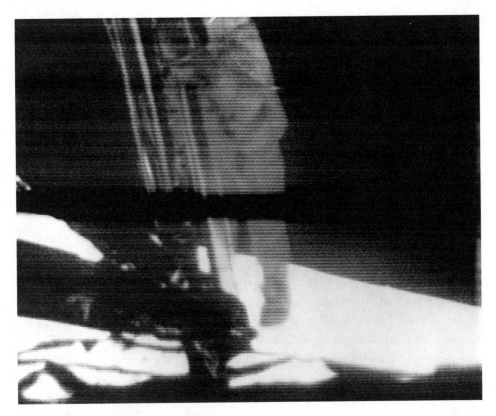

Figure 5.15
Televised image of Astronaut Neil Armstrong stepping onto the Moon. Courtesy of NASA Marshall Space Flight Center.

exploration subsided and Congress funded a more modest U.S. space program after Apollo.[56] Astronauts occupied a temporary space station, Skylab, in Earth orbit from 1973 to 1974, followed after 1981 by voyages into orbit of a reusable craft, the Space Shuttle, which divided its time between defense-related missions and civilian scientific ones.[57] The astronaut corps became more diverse in the 1970s and 1980s, as civilian scientists and mission specialists joined military pilots.[58] The Shuttle program ended in 2011, though, and for a time the United States depended ironically on Russia, the losing side in the race to the Moon, for rockets to access the International Space Station, a permanent station built in Earth orbit in 1998–2000.[59]

The Apollo program required the coordinated work of half a million people and a budget of $20 billion (equivalent to about $180 billion in 2010).[60] About half of this effort went into the rocket boosters. The federal agency in charge, NASA, was the largest public civilian engineering organization to be created in the United States in peacetime since the Tennessee Valley Authority of the 1930s. Like the TVA, NASA in its early years had three leaders: James Webb, former federal budget director, served as administrator from 1961 to 1968; Hugh Dryden, former head of NASA's predecessor body, NACA, was his deputy to 1965; and the engineer Robert Seamans was associate director. Unlike the three heads of the early TVA, the three NASA leaders worked well together. Just as Hyman Rickover did in his organization of the nuclear navy, the NASA leaders also enforced a policy of carefully documenting every decision and every detail so that everyone knew what they were responsible for doing and leaders at every level could keep track of how all of the work connected together.[61]

Unlike the TVA, which did its own design and construction, NASA relied on outside private contractors to do most of the design and assembly for the Apollo program. The task of reaching the Moon was too large and complex to create a new organization with the necessary people and skills. NASA confined itself to overseeing the contractors and intervening when necessary to fix problems and keep the work on schedule. The Apollo program was still a radical achievement involving large private firms and many smaller ones, working together under centralized public direction.[62]

The lunar landing benefited from scientific research of the upper atmosphere in the late 1950s and from the solution of other scientific questions as needed, but no radically new scientific breakthrough initiated the Apollo program. The principle of rocket propulsion was Newton's third law of motion, and the other basic ideas of rocketry were engineering ideas or formulas worked out by Robert Goddard and others earlier in the twentieth century. Unlike earlier engineering achievements, though, Apollo was the result not of one or two guiding intellects but of many minds brought together.[63]

Mission Control director Gene Kranz captured the determination of the enterprise in his famous words during the Apollo 13 crisis: "Failure is not an option."[64] After Apollo, however, the American space program settled into a more routine and varied set of missions and lost some of its earlier vigilance. In 1986, a rubber ring failed on one of the booster rockets for the space shuttle *Challenger*, causing the rocket to explode as it rose into the sky, killing the crew; and in 2003, the space shuttle *Columbia* broke apart on reentry when heat-shielding tiles broke away from one of its wings. In the first case, management proceeded with the flight despite a warning from an engineer, and in the second case, tile shedding had occurred before and was deemed an acceptable risk.[65]

The lasting impact and benefits of the Apollo program have been debated since the Moon landing in 1969.[66] Neil Armstrong, himself an engineer, rated space exploration the greatest engineering achievement of the twentieth century but only twelfth in social benefit to the world.[67] In retrospect, one aspect seems clear. The space program of the 1960s was unique, not because of the new industry that it helped to establish, although a new industry did emerge around satellite communications and commercial rocket launches. It was unique because of its purpose. Reaching the Moon first was an engineering achievement that could happen only once. The goal was a public one that inspired a unifying spirit of sacrifice, ingenuity, and care among those involved that only an extraordinary national mission focused on a singular moment could evoke.

The lunar landings required new electronics as well as new rocketry. If the exploration of outer space marked the farthest that modern engineering could venture outward, the advances in electronics that came after 1945 brought an equally radical venture inward, through miniaturization in the scales at which engineers could design and make things. The results of this miniaturization were vital to the Apollo missions and would eventually change life and work, not only in America but in the world.

6 Bardeen and the Transistor

The second half of the twentieth century was a second age of electronics. The first began with the invention in 1906 of a vacuum-tube amplifier, the triode, and the advances in electronic circuit design that followed, which made possible long-distance telephony, radio transmission and reception, and then television. The second age began in 1947 with the invention of the transistor, a small device that could perform the functions of a vacuum-tube amplifier without having to enclose a vacuum.

During the 1930s, advances in quantum physics gave engineers and scientists more knowledge about certain elements called semiconductors. In 1945, a scientist at the Bell Telephone Laboratories, William Shockley, attempted to apply these findings in an experiment to see whether a semiconductor could be used to amplify an electric current, dispensing with a vacuum tube. The experiment failed. Another Bell scientist, John Bardeen, soon discovered that properties of the semiconductor surface had a blocking effect. With the help of another scientist, Walter Brattain, Bardeen found a way to overcome this effect. In late 1947, Bardeen and Brattain then designed a circuit, the transistor, that successfully amplified a current through a semiconductor. Once he saw how it worked, Shockley invented a better version, and in the early 1950s manufacturers improved it further to produce commercial transistors.

Accounts of the transistor have stressed the role of new physics as a stimulus to its invention. However, the breakthrough occurred after Bell researchers found that the science in 1945 wasn't enough for simple application. Bardeen and Brattain needed to investigate the natural properties of semiconductors more deeply, and what they learned was counterintuitive to what the science had led them to expect. To focus their research, they also needed a prior engineering objective, and their work was possible only because of engineering advances during World War II in the refining of semiconductors. The transistor was an example not of engineering as simply applied science but of how engineering and science were equally basic to an innovation.

The Vacuum Tube and Its Limits

In 1899, the Bell Telephone Company changed its formal name to the American Telephone and Telegraph Company, although it was known for short as the Bell System. By then, telephone calls could reach halfway across the United States. The technology of the American telephone network could not extend the range of long-distance calls much farther. Early radio equipment could soon transmit telegraph signals over several thousand miles but could carry voice only over short distances.[1]

Two new devices overcame these limitations. In 1904, an English engineer, John Ambrose Fleming, inserted a metal plate near the filament inside a light bulb. Electrons had a negative charge, and he found that when the plate had a positive charge, the heated filament emitted electrons that flowed to the plate through the vacuum inside the bulb, causing a current to flow in a wire leading away from the plate. If the plate had a negative charge, no electrons flowed through the vacuum. Most of the electricity used in homes and workplaces came from generating stations in a form known as *alternating current*, in which the current reverses direction many times per second. The two directions are positive and negative. In a Fleming device, this kind of current gave the plate an alternating positive or negative charge. When positive, the plate "rectified" the alternating current into *direct current*, or current flowing only in one direction. The Fleming bulb was known as a *diode* rectifier for its two elements, the filament and the plate.[2]

Two years later in 1906, an American engineer, Lee de Forest, found a way to make a diode into an amplifier of electric current. By inserting a small wire grid between the filament and plate of a diode, and charging the grid with a small alternating current, de Forest attracted more electrons from the filament to the plate when the grid charge was positive. The result was to amplify the current leaving the plate. De Forest called his device an "audion" but engineers soon renamed it a *triode* (box 6.1). Vacuum bulbs to rectify, or rectify and amplify, electric current came to be known among American electrical engineers as vacuum tubes.[3]

The Bell System acquired the patent to the de Forest triode, and Bell engineers improved it to amplify long-distance telephone calls, making transcontinental phone connections possible by 1915.[4] During the 1910s and 1920s, the American engineer Edwin Howard Armstrong made a number of innovations in the design of radio receivers that employed both diodes and triodes. His work improved amplification and made it practical for radio sets to receive and play clear and audible sound. Larger triodes also made it practical to broadcast radio waves over longer distances.[5]

By the 1930s, with the help of the new tubes, telephony and radio had changed dramatically the way Americans communicated with each other and shared news and

Box 6.1
Vacuum-Tube Electronics

In a household *electric* circuit, a current flows or does not flow with the flip of a mechanical switch. An example is an electric light that switches on and off.

In an *electronic* circuit, a second flow of electric charge controls a first flow of charge. When electricity is turned on, it is available to both flows. The second flow of charge can switch the first flow on or off, and can amplify it. In early electronic devices, the flows of charge occurred in a glass bulb or tube that enclosed a vacuum.

Two of these devices, the diode and triode, modified the form of electricity most commonly in use, *alternating current*, in which the current reverses direction many times per second. The two directions are denoted as positive or negative. The other form of electricity, *direct current*, flows in only one direction. Blocking the negative direction of flow in an alternating current is a way to produce a direct current.[a]

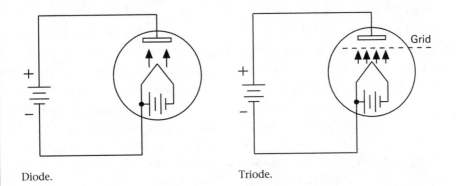

Diode.   Triode.

In a *diode* (*above, left*), a plate attached to the positive (+) end of a battery is positively charged and attracts negatively charged electrons (the arrows) emitted by a light bulb filament attached to a separate battery. However, if the plate attaches to the negative end of its battery (–), the negative charges repel each other and no electrons flow through the vacuum. Connected to an alternating current instead of a battery, the plate only attracts electrons when the current is positive. As a result, the device can "rectify" an alternating current into a direct current in the plate.

In a *triode* (*above, right*), a grid or screen (- - -) with a small alternating current stands between the filament and plate of a diode (source of grid current not shown). When the grid is positive, it increases the flow of electrons reaching a positively charged plate. These amplify as well as rectify the plate current. When the grid is negative, electrons do not flow.

Source: Abraham Marcus and William Marcus, *Elements of Radio* (New York: Prentice-Hall, 1943), 99, 107–110.
Note: (a) The flicker from alternating current, or the blocking for direct current, isn't noticeable.

entertainment. However, vacuum tubes had drawbacks that limited the further progress of electronics using them. The tubes could not be made much smaller and they
generated heat and often burned out. Their failure rate made using them in large assemblies especially difficult. These limitations were a particular challenge to telephony. In
addition to needing amplification, telephones were connected to each other through
exchanges that relied on mechanical switches. The growing number of calls would
eventually exceed the speed at which the switches could pass them through. Vacuum
tubes could work as faster switches, but their drawbacks made them unreliable.[6]

In 1925, American Telephone and Telegraph organized the Bell Telephone Laboratories in New York City to meet its growing technical needs. During the 1930s, the
director of research at Bell Labs, Mervin J. Kelly, decided that the telephone network
would need someday to replace both vacuum tubes and mechanical switches. Kelly and
others at Bell Labs became aware of advances in modern physics that gave researchers a
new understanding of conductivity in solid materials. The new physics encouraged the
idea that it might be possible for a "solid-state" device—a device made of solid material
without a glass-enclosed vacuum—to perform the functions of a vacuum tube.

Scientists knew that metals, such as copper, could conduct electricity well, while
other materials, such as rubber, were insulators. The new science, known as quantum
physics, explained that in highly conductive materials, a small number of electrons
detached from their original atoms and were able to conduct electric charge. Electron
vacancies in atoms, called "holes," could also conduct charge. In good insulators, on
the other hand, there were almost no free electrons or holes with the ability to conduct
charge. An intermediate class of materials, known as *semiconductors*, served as insulators but could conduct electricity when heated (box 6.2). In their natural state, these
materials were unreliable, but this would soon change.[7]

Radar and the Refining of Silicon

During World War II, America called on its trained engineers and scientists to develop
new weapons. In addition to nuclear bombs, the war introduced devices that employed
radio waves to detect objects at a distance, and the wartime research that led to better
radio detection made it possible to develop a solid-state amplifier after the war.

Following the demonstration of wireless telegraphy in the early 1900s, engineers in
many countries realized that radio waves might also be used for detection at a distance.
Radio-wave detection became practical in Britain in the 1930s, when a team led by
Robert Watson-Watt, a descendant of the steam engine innovator James Watt, devised a
system that could detect formations of enemy airplanes in flight from a chain of coastal

Box 6.2

Semiconductors

Conductors, Insulators, and Semiconductors

During the 1930s, the new science of quantum physics gave a better grasp of how conducting and insulating materials worked. Scientists found that electrons surround atomic nuclei in concentric bands, and some electrons in the outermost, or "valence," band are able to break free of their nuclei. In good *conductors*, these free electrons can move into an adjacent zone, called the conduction band, where they can conduct electric charge easily. In materials that are good *insulators*, very few free electrons can enter this zone because of a gap that separates it from the valence band.[a]

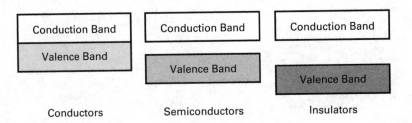

| Conductors | Semiconductors | Insulators |

Certain elements, called *semiconductors*, normally function as insulators. In these elements, though, a smaller gap separates free electrons in the valence band from the conduction band. When the semiconductor is heated (or specially treated), free electrons can enter the conduction band and conduct charge. Freed electrons leave vacancies or "holes" in their atoms that can also conduct charge.

The two most useful semiconductors have been the elements silicon and germanium. In their natural form, however, these elements are erratic conductors because of impurities in the material. Semiconductors could not be useful until engineers learned how to purify and further treat them to improve their conductivity.

Treating Semiconductors

During World War II, American and British engineers needed reliable silicon diodes for radar. By learning how to purify the silicon, and by then adding a small amount of another element (called "doping"), engineers created an excess of free electrons or holes to make silicon more reliably conductive.

Silicon has four electrons in its valence band. Adding phosphorus, with five valence electrons, increased the free electrons, while adding boron, with only three valence electrons, in effect created holes. Semiconductors came to be known as "n-type," when they had an excess of (negatively charged) electrons, or "p-type," when they had an excess of (positively charged) holes.

Source: G. L. Pearson and W. H. Brattain, "History of Semiconductor Research," *Proceedings of the Institute of Radio Engineers* 43, no. 12 (December 1955): 1794–1806. *Note:* (a) Drawings shaded for visual contrast only.

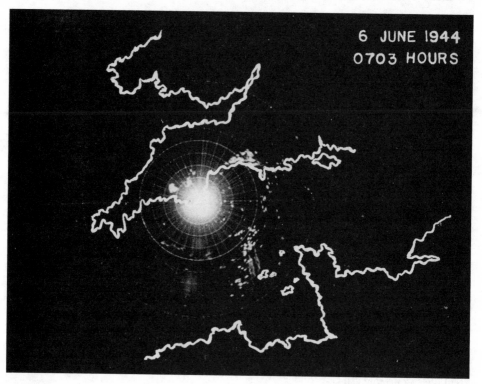

Figure 6.1
Radar image of the English Channel on the morning of June 6, 1944 (D-Day) during World War
II. Coastlines appear to be drawn in. Courtesy of U.S. National Archives.

stations. In the summer of 1940, Watson-Watt's system gave the British Royal Air Force
critical early warning of attacking German warplanes and helped the British defeat the
Germans in the air. The British called radio-wave detection RDF (for Range and Direction Finding). In the United States, the technology came to be known as *radar* (short
for "radio detection and ranging") (figure 6.1).[8]

After the fall of France in June 1940, President Franklin Roosevelt appointed a research
committee to improve American weapons in the event that the United States had to
enter the war. As committee chairman, the President chose Vannevar Bush (1890–1974),
an MIT engineer serving as director of the Carnegie Institution for Science in Washington, DC (figure 6.2). Bush formed a laboratory at MIT to develop better radar.[9]

Radar worked by emitting radio waves and then detecting their backward flow when
they bounced off a target. Using rectifiers, radar equipment could display the returning

Figure 6.2
Vannevar Bush, director of the U.S. Office of Scientific Research and Development during World War II. Courtesy of U.S. Library of Congress.

signals on screens, but British radar was not precise enough to detect individual airplanes from a long distance, only formations. To give more precise information, radar systems needed to emit radio waves at microwave frequencies, which were too high for the vacuum tubes that Watson-Watt used to receive and rectify signals. The British soon devised a machine (the cavity magnetron) to generate microwave frequencies without vacuum tubes, and a researcher at Bell Labs discovered that a semiconductor could rectify these high-frequency waves for display on radar screens. However, Bell researchers found that natural semiconductors were erratic in their conductivity owing to the presence of impurities. If the impurities could be removed and other elements added back in very small controlled amounts (a process called *doping*), a semiconductor could be made reliable enough for use in microwave radar.[10]

The easiest semiconductors to purify were the elements germanium and silicon. With existing methods, neither could be refined to more than 99 percent purity, which was not good enough. A solution came from the DuPont chemical company in Delaware. The war cut off DuPont from its supply in India of titanium, a mineral used in the coloring of white paint, and silicon was a good candidate to replace titanium if it could be purified. C. Marcus Olson, a chemist with DuPont, combined a form of silicon, silicon tetrachloride, with zinc, by melting 99.999 percent pure zinc and mixing its vapor with the vapor of melted silicon tetrachloride. The process produced zinc chloride and excess silicon chloride, both of which Olson was able to remove, leaving needles of silicon with a purity of 99.999 percent.[11]

The Radiation Laboratory at MIT obtained the purified silicon, and by adding back minute amounts of other elements (phosphorus or boron) to promote conductivity, the laboratory was able to design the diode rectifiers needed to operate high-frequency radar systems after America entered World War II in 1941. By the end of the war, a team at Purdue University had found a way to purify germanium as well. Having purer forms of silicon and germanium, and having the ability to enhance their conductivity through controlled doping, proved crucial to the postwar development of the transistor.[12]

The Solid-State Conundrum

After the war, the Bell Telephone Laboratories moved from New York City to a new and more secluded campus in Murray Hill, New Jersey, where their engineers and scientists resumed civilian work (figure 6.3). The most important need in the postwar telephone industry was to find a better way to amplify and switch the electric currents that carried telephone calls. If a solid-state device could perform these tasks, it could take the place of vacuum-tube triodes. Silicon diode rectifiers were solid-state devices that had proved themselves in radar. As the war reached its end, Mervin Kelly, the director of Bell research, began a program to develop a solid-state triode.[13]

To lead the program, Kelly appointed a quantum physicist, William Shockley (1910–1989), who had grown up in Palo Alto, California, where his father was a mining engineer and his mother was a surveyor. The younger Shockley graduated from the California Institute of Technology in Pasadena in 1932 and earned a doctorate in physics from the Massachusetts Institute of Technology four years later. Kelly hired him in 1936 to do solid-state research but most of his work in the late 1930s was to meet other technical needs. During the war, Shockley went to Washington, where he worked to improve the effectiveness of strategic bombing, and he helped train pilots in the use of

Figure 6.3
Entrance to the Bell Telephone Laboratories in Murray Hill, New Jersey, in the late 1940s. Courtesy of U.S. Library of Congress. Gottscho-Schleisner, Inc., photographer.

new radar-controlled bombsights. He returned to Bell Labs part-time in the spring and summer of 1945 before resuming full-time work in the autumn.[14]

On his return in the spring of 1945, Shockley began experiments to see if a semiconductor could be made to amplify an electric current. During the war, engineers had begun to classify the semiconducting elements, silicon and germanium, according to the way they conducted electric charge after doping. A semiconductor with excess electrons came to be called "n-type," so-named because electrons have a negative charge, while a semiconductor with excess holes came to be called "p-type" because the holes acted as if they had a positive charge. Free electrons were the "majority" carriers of charge in n-type material, while holes were the majority carriers in p-type material. Shockley decided to work with n-type silicon.[15]

In a vacuum-tube triode, a positively charged grid amplified the flow of negatively charged electrons from a filament, through the grid, to a positively charged plate. Shockley designed an experiment without a vacuum, in which he hoped to see whether

Box 6.3

Shockley's Field-Effect Experiments

By 1945, semiconductors were reliable enough to investigate whether they could perform the functions of a triode.

William Shockley, a physicist at Bell Laboratories, carried out experiments in the spring and summer of 1945, in which he positioned a metal plate close to a strip of n-type silicon. He believed that when the plate was positively charged (+), it would attract negatively charged (–) electrons to the surface of the silicon and would amplify an electric current going through the surface.

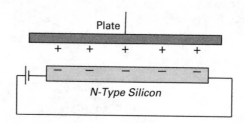

Since amplification was to be achieved by applying an electric field across the silicon, it was termed a "field effect" experiment. After many trials, though, Shockley found negligible amplification. He gave the problem to John Bardeen, a theoretical physicist, and Walter Brattain, an experimental physicist, to solve.

Source: William Shockley, "The Path to the Conception of the Junction Transistor," *IEEE Transactions on Electron Devices* ED-23, no. 7 (July 1976): 604–605.

a plate with positive charge, positioned close to a strip of n-type silicon, might attract electrons inside the silicon to the silicon surface. He reasoned that these electrons would amplify another electric charge flowing across the surface. The amplification he hoped to achieve would be a "field effect," since the plate would create an electric field over the silicon below. Shockley conducted experiments in which he ran an electric current through a thin strip of n-type silicon. Close to the strip, separated by a small amount of air, he positioned a plate and gave it a positive charge. However, in several tests, Shockley failed to detect more than minuscule amplification (box 6.3).[16]

Bardeen, Brattain, and Breakthrough

On his return to Bell Labs full-time in September, Shockley took charge of a small group of physicists, chemists, and engineers formed by Mervin Kelly to do research in solid-state physics. Shockley also oversaw a small subgroup that would focus on semi-conductors. Kelly assigned two scientists to Shockley's subgroup who would play key roles. The first was Walter Brattain (1902–1987), who grew up on a farm in the state of Washington, graduated from Whitman College in 1924, and earned a master's degree from the University of Washington and a doctorate in physics from the University of Minnesota in 1929. He had worked for Bell Laboratories in New York in the 1930s, and he was returning from war work for the navy.[17]

The second was John Bardeen (1908–1991), who grew up in Madison, Wisconsin. Bardeen's father was the first dean of the medical school at the University of Wisconsin and his mother was an interior decorator. Mathematically gifted, John earned a master's degree in electrical engineering from the University of Wisconsin in 1928; he found work privately as a geophysicist until 1933. In that year he began to study for a doctorate in mathematical physics at Princeton University, where he played cards with Robert Brattain, Walter's brother, and met Walter, who came down for the games. John Bardeen's interest was quantum theory as it applied to metals, and he received his degree in 1936. After a fellowship at Harvard, he began teaching physics at the University of Minnesota. During the war, he helped develop new minesweeping measures for the Navy. The University of Minnesota did not offer Bardeen a better salary if he returned, so after the war he accepted an offer from Mervin Kelly to join Bell Labs.[18]

Bardeen, Brattain, and Shockley (figure 6.4) all had a long-standing interest in electronics from building amateur radio sets in their youth. Each also brought special abilities to their work at Bell Labs. Shockley stimulated others with a constant flow of ideas and an ability to explain complex ones. Brattain was more easygoing and found challenge in testing insights experimentally. Bardeen was a more self-effacing theorist and wanted to understand why ideas didn't work as expected.

Taking note of Shockley's experiments the previous spring, the semiconductor group made two strategic decisions that autumn with Kelly's support. First, they agreed that the best materials for a solid-state triode were the semiconductors silicon and germanium, purified and then slightly doped with other elements to make them more conductive. Second, with the failure of Shockley's experiments in mind, the group decided that the natural properties of silicon and germanium needed deeper scientific investigation before further engineering work could proceed.[19]

Over the winter of 1945–46, Bardeen pondered why Shockley had failed to amplify a current. Bardeen knew from his own doctoral research that the surfaces of certain metals could differ electrically from the interior of the metals. A series of experiments by Brattain confirmed that a similar difference was true of semiconductors. In March

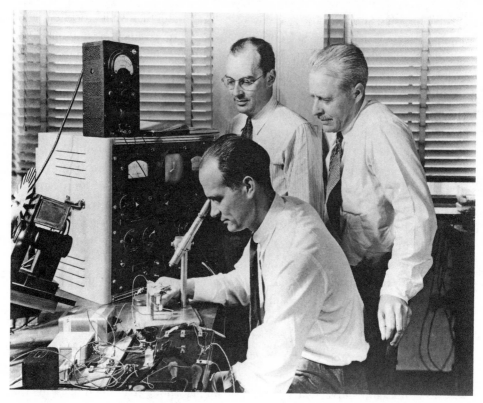

Figure 6.4
John Bardeen (*standing, left*), Walter Brattain (*standing, right*), and William Shockley (*seated*) in 1948. Courtesy of AT&T Archives and History Center.

1946, Bardeen concluded that Shockley had failed because a thin layer of excess electrons adhered to the surface of the n-type silicon and a thin layer of excess holes collected beneath it. The density of these states slowed the mobility of electrons at the surface and prevented the positive charge above from amplifying a current below.[20]

Over the next year and a half, Bardeen and Brattain tried to overcome this barrier. Shockley's own focus shifted to other work, leaving the two to solve the problem. Bardeen reflected on each step of the research as a theorist while Brattain designed and carried out experiments to test each step they took. After many frustrating setbacks, on November 17, 1947, they finally achieved a breakthrough. Instead of using a charged plate separated from silicon by open space, Brattain placed a metal contact on a spot of conductive fluid (called an *electrolyte*) that touched the silicon. By then applying a charge to the metal contact, he was able to reduce the density of the energy states at the silicon surface sufficiently to overcome their blocking effect.

The next step was to see if a charge applied in this way could amplify a flow of charge below the surface. Bardeen and Brattain switched from silicon to a form of n-type germanium, which they thought would produce amplification more easily, and they replaced a fluid electrolyte with a solid electrolyte, gold. The next step was to apply a positive charge to attract electrons in the germanium to a position close to the surface, where they could amplify a current going through them. The charge instead pulled electrons into the contact and attracted holes into the desired position in the germanium below (the holes, having a positive charge, moved to a position outside the positive contact because like charges repelled each other). Another experiment to reduce this effect only increased it. Bardeen soon realized that the holes might perform the desired amplification. This was contrary to the expectation that electrons, the majority carriers of charge in an n-type semiconductor, should perform the amplification.[21]

On December 16, Bardeen and Brattain placed two gold metal contacts close together on the tip of a plastic wedge touching a piece of n-type germanium below. Each contact had an electric charge, one positive and one negative. The positive contact, called the *emitter*, attracted electrons into the contact and increased the number of holes in the germanium just below. The holes were close enough to the other contact, called the *collector*, to amplify a current through the other contact. Bardeen and Brattain refined the experiment and used the collector current to amplify sound through headphones in a demonstration to higher executives on December 23. A colleague gave the device a name, transfer resistor, or *transistor* (box 6.4; figure 6.5), for its ability to modify the conductivity (and thus the resistance) of the semiconductor.[22]

Bardeen and Brattain were not the only ones studying semiconductors at the time. It was the wartime work of the researchers at Purdue University on germanium that convinced Bardeen and Brattain that germanium would be easier to work with in the last stage of their transistor effort. In early 1947, almost a year before the breakthrough at Bell Labs, a Purdue graduate student noticed that applying a positive voltage increased the conductivity of n-type germanium. The Purdue group didn't go on to invent the transistor, though, because they weren't looking for such a device. After the war, the Purdue researchers resumed the traditional academic mission of doing basic scientific research. Their goal was to obtain more knowledge of the natural properties in semiconductors rather than serve an immediate practical purpose. Although Bardeen and Brattain also conducted new research into natural properties, a practical goal focused their work and enabled them to recognize the breakthrough when it came.[23]

Box 6.4

Breakthrough to the Transistor

After the failure of Shockley's 1945 experiments, John Bardeen reflected on the problem. In March 1946 he recognized that electrons were trapped on the surface of the semiconductor by energy states (densities) that prevented amplification. After many setbacks, Bardeen and his colleague, Walter Brattain, found a way in November 1947 to penetrate the surface energy states by applying an electric charge to a metal contact through a conductive electrolyte (a gold spot) placed on an n-type semiconductor.

The two researchers then expected a positive charge in the contact to attract free electrons in the semiconductor toward the surface, where they would amplify a current leading away from the semiconductor. However, a positive charge instead attracted the electrons into the contact, concentrating holes in the semiconductor below because the holes also had a positive charge and like charges resisted each other. The two researchers realized that they could use the holes, instead of free electrons, to amplify a current through the semiconductor, since holes could also amplify a current.

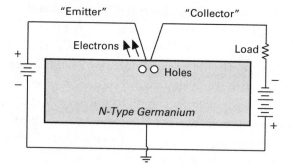

Diagram of the experimental transistor circuit.

Bardeen and Brattain demonstrated amplification of sound in headphones (the "load" shown in the above diagram) in December 1947. By placing a positively charged contact (called the "emitter") close to a negative contact (called the "collector"), both on the surface of an n-type piece of germanium, the positive contact attracted electrons into itself and brought holes close to the negative contact. The holes amplified a current through the collector leading to the headphones. The minority carriers of charge in n-type material thus became the means to the desired amplification.

Source: J. Bardeen and W. H. Brattain, "The Transistor, A Semiconductor Triode," *Physical Review* 74, no. 2 (July 15, 1948): 230–231.

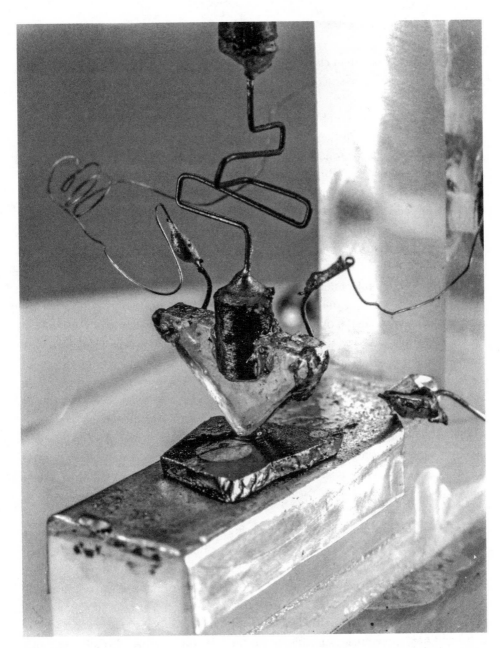

Figure 6.5
The experimental transistor demonstrated at Bell Labs in December 1947. Two gold contacts adhered to the converging sides of a triangular plastic wedge, leaving a small gap at the tip touching a flat piece of n-type germanium. Courtesy of AT&T History Center and Archives.

The Rise of Transistor Electronics

After filing a series of patent claims, Bell Labs announced the transistor to the world in June 1948. Despite successful demonstrations, however, there were serious drawbacks to the "point-contact" transistor of Bardeen and Brattain. The original device was a laboratory experiment assembled with improvised parts. It was not a design suitable for manufacturing, and further study showed that the point contacts did not perform as reliably as hoped. A personal conflict also arose. Although Shockley gave occasional advice to Bardeen and Brattain during their research, he did not participate in the work that led to the breakthrough, and Bell attorneys found that his own idea of a field effect had been anticipated by a 1930 patent claim. When the Bell attorneys decided not to include Shockley's name on the patent application, he was embittered and resolved to make a better transistor himself, without telling Bardeen or Brattain.[24]

Over the winter of 1947–48, Shockley conceived a different way to achieve amplification through a semiconductor. Instead of having two contacts on one surface, he proposed inserting a thin slice of p-type germanium between two larger sections of n-type. The n-type material on one side served as the emitter and the n-type material on the other side served as the collector. In this device, a small current in the p-type material, called the base, attracted electrons from the emitter side and amplified them in the collector. Turning the base current on and off could also make the device work as a switch. Shockley called his device a "junction" transistor, and an experiment in January 1948 gave evidence that it could work (box 6.5).[25] The junction device needed further improvement that Gordon Teal, a chemist at Bell Labs, achieved by making germanium more uniform in its composition.[26] In 1951, Bell Telephone began licensing the transistor, and most early licensees chose to make junction devices, using germanium. A key improvement was the addition of alloys by RCA and General Electric to make them easier to manufacture.[27]

The first use of the transistor was in hearing aids. But the U.S. armed forces soon became the largest buyer of transistors for use in military electronics, and military orders in the early 1950s were decisive in enabling manufacturers to increase production and bring down the cost per unit. In 1954, Texas Instruments expanded the civilian market by making the first hand-held transistor radios in partnership with a small electronics firm. The Regency TR-1 radio (figure 6.6) sold for $49.95 (about $400 in 2010) and became popular. A Japanese firm, now known as Sony, also took a license to produce transistor radios and entered the American market in 1957. A worldwide market came into being as the price of transistor radios fell in the 1960s (figure 6.7).[28]

Box 6.5

The Junction Transistor

The demonstration of a point-contact transistor proved that a solid-state triode was possible. However, in January 1948, William Shockley invented a more reliable device, the junction transistor, in which a section of n-type or p-type semiconductor, called the *base*, stood between an emitter and collector of opposite type material.

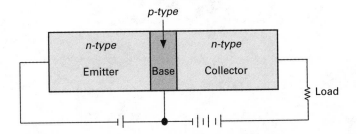

A control current entering the base attracted an opposite current in the emitter and amplified it in the collector. In the drawing above, a positive charge in the base would attract negatively charged electrons from the emitter and would amplify them in the collector. The junction device employed two separate materials, n-type and p-type, instead of one material (n-type or p-type) as in the point-contact device, to supply electrons and holes.

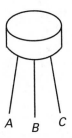

A typical junction transistor of the 1950s resembled the drawing above, in which the emitter current went into the device through the lead *A*, the base charge went in through the lead *B*, and the collector current went out through the lead *C*.

Source: William Shockley, "The Path to the Conception of the Junction Transistor," *IEEE Transactions on Electron Devices* ED-23, no. 7 (July 1976): 612–613.

TR-1

THE FIRST
TRANSISTOR
RADIO RECEIVER

TECHNICAL DATA
AND
SERVICE NOTES

REGENCY DIV. I.D.E.A. INC.
7900 Pendleton Pike, Indianapolis 26, Ind.

25 CENTS

Figure 6.6
Regency TR-1 owner's manual title page. Courtesy of Don Pies.

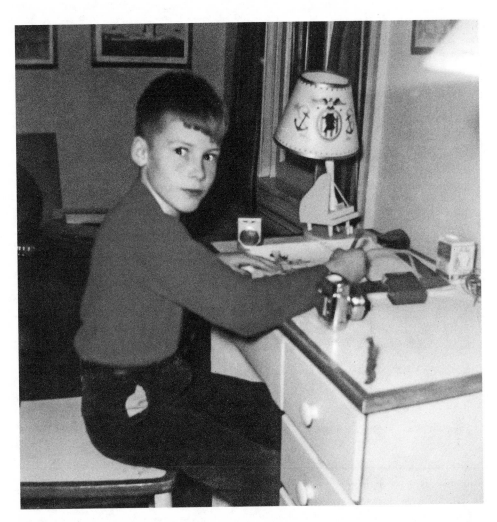

Figure 6.7
The author, David P. Billington, Jr., with a Hilton 6 transistor radio (*to his left*). From a Billington family album.

Box 6.6

The MOSFET Transistor

Point-contact and junction transistors were known as *bipolar* devices, because they employed two charge carriers, electrons and holes. In 1959, M. M. Atalla and Dawon Kahng at Bell Labs created an improved transistor that used just a single charge carrier. The new device became known as a metal oxide semiconductor field effect transistor, or MOSFET for short. The emitter and collector were now called the *source* and the *drain*. The base or gate had a thin insulating layer of silicon dioxide (not shown) under it.

The gate worked with voltage, a force or pressure in electricity that makes it move. With n-type material under the source and the drain, and p-type material between them, a positive voltage on the gate caused a field underneath that could amplify (or turn on and off) a flow of electrons in the p-type material between the source and the drain. Holes weren't needed to amplify the flow. The device also worked with p-type material below the source and the drain and n-type material between them. In this case, with a negative gate voltage, the flow consisted of holes with no electrons involved.

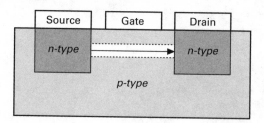

Diagram of a field-effect transistor.

Eliminating the second charge carrier made the transistor more efficient. Field-effect transistors replaced junction transistors in the 1970s for most uses in electronics.

Source: Bernard Grob, *Basic Electronics*, 5th ed. (New York: McGraw-Hill, 1984), 585–586. Some details are omitted from the diagram.

The field effect that Shockley failed to achieve eventually worked, although not in the way that he envisioned in 1945. In 1959, two engineers at Bell Labs, Mohamed M. Atalla and Dawon Kahng invented a new kind of transistor, known as a MOSFET, short for metal oxide semiconductor field effect transistor (box 6.6) On a top surface, over an insulating layer of silicon dioxide, a base contact now called the *gate* stood over a region of p-type material below it. On either side were contacts for an emitter current, now called the *source*, and a collector current, now called the *drain*. Below each of the latter two stood a region of n-type material. When the gate received electricity, its moving force or voltage created a field below that modified a flow of electrons between the source and the drain, amplifying the flow, or turning it on and off as a switch. The device could also work with the gate over n-type material and p-type material under the source and the drain, in which case holes flowed from one to the other. The device depended on a single carrier of charge, not on the action of two carriers (electrons and holes) as in point-contact and junction transistors. The application of voltage had a field effect, and most later transistors were devices of this kind.[29]

John Bardeen, Walter Brattain, and William Shockley shared the 1956 Nobel Prize in Physics for their transistor work. However, Shockley distanced himself from the other two after the winter of 1947–48 and grew more difficult for Bardeen and Brattain to work under. Bell Labs did not want to support Bardeen in the more theoretical work that he wanted now to do, and in 1951 he accepted a faculty position with the University of Illinois at Urbana-Champaign. Brattain moved to another group at Bell Labs and retired in 1967 to accept a teaching position at Whitman College.

Shockley quit Bell Labs in 1955 to found a company in Mountain View, California, near his hometown of Palo Alto, to design and produce transistors, but his business failed and he joined the faculty at Stanford University. In later life, he became obsessed with the idea that human racial groups were not equal in their intelligence and that people of African descent were inferior to Europeans. In holding such ideas, he was not unlike Henry Ford, an innovator who in the 1920s gave expression to anti-Semitic views. Unlike Ford, who offered a public apology later, Shockley held to his views.[30]

Bardeen in later life lived quietly at the University of Illinois and continued to do groundbreaking work. He shared a second Nobel Prize in Physics in 1972 for research on superconductivity, becoming the first person to win the prize in physics twice. His work was of value to both engineering and science, and at Illinois, he was a professor in both electrical engineering and physics.[31]

Science and Engineering in the Transistor

After 1945, America began to embrace the idea that modern technical innovation came by applying the work of basic science, or unpressured inquiry to explore nature for its own sake. The transistor has been cited since then as an example, if not the premier example, of an innovation that owed its inspiration to such inquiry.

The idea of engineering as applied science received influential support in 1945 from Vannevar Bush, the director of wartime research in the United States. In the autumn of 1944, President Roosevelt asked Bush to report on scientific needs in the coming postwar era. In his report, delivered to President Truman in July 1945 and published as *Science, the Endless Frontier*, Bush called for peacetime federal support for scientific research in the nation's universities and called for federal scholarships to give many more Americans advanced training in the sciences. But he also made a sweeping historical assertion: that the advances of modern technology originated in basic science. "Basic research is performed without thought of practical ends," he wrote. "It results in general knowledge and an understanding of nature and its laws. . . . New products and new processes . . . are founded on new principles and new conceptions, which in turn are painstakingly developed by research in the purest realms of science."[32]

In addition to the wartime work on radar, which made use of the scientific discovery of radio waves in the 1880s, Bush had in mind medical research that led to the wartime development of the antibiotic penicillin.[33] A basic kind of study also contributed to the invention of the point-contact transistor. Mervin Kelly insulated Shockley, Bardeen, and Brattain from the normal work of telephone research, which was to improve service and solve more mundane technical problems. However, this was not a mandate to do science for its own sake in the sense that Bush had argued. Bardeen and Brattain had an engineering mission to develop a solid-state triode. The Bell researchers were given the freedom to study the natural properties of semiconductors, but they worked with a clear and urgent engineering objective in mind.

In his 1945 experiments, it was Shockley who tried to act as an applied scientist. Quantum theory gave him the principles he needed to produce amplification, or so he thought, and the problem seemed a matter of application, of putting pieces together and watching them work. When his experiments failed, instead of questioning the science more deeply himself, he left the problem to Bardeen and Brattain. After months of study, the two researchers finally succeeded by using holes to amplify a current. Once Shockley realized that "minority carriers" of charge were the key to transistor operation, he quickly developed a better transistor of his own, but his original experiments in 1945 were much closer to the image of applied science. The term "applied science" continues to be used alongside engineering to mean what engineers do, but engineers understand it to mean the kind of work done by Bardeen and Brattain, and others like Edison before them.

The triode principle went back to 1906, before quantum physics existed. Quantum physics made a solid-state triode imaginable in the 1930s but did not provide the knowledge sufficient to create one, and a semiconductor triode would not have been possible without wartime engineering advances in semiconductor refining. Bardeen's success in 1947 involved engineering a desired effect as much as discovering how the natural materials behaved. Engineering and science were thus equally "basic" to the innovation of the transistor. Science and engineering did not derive one from the other, and the transistor innovators had to be able to do both.[34]

Bell Labs continued to produce important advances in engineering and science, although none was as dramatic in its impact as the transistor.[35] The device proved that working in an age of greater intellectual and bureaucratic specialization was not a barrier to radical innovation, but neither was it a guarantee. Bardeen needed colleagues, the support of a technically proficient organization, and his own prior training to succeed. However, the original insight was not the automatic result of such benefits; other firms and researchers had them too. What Bardeen added was an ability to use these advantages to think radically and independently, as both a scientist and an engineer.

In the decade that followed its innovation, limits to transistor miniaturization threatened to bring the advance of electronics to a standstill. Two engineers, Jack Kilby and Robert Noyce, working in different parts of the country for small firms, found a solution that eluded larger and more established electronics manufacturers working for the federal government. The solution, the integrated circuit, or microchip, would shape America's course for the rest of the century and beyond.

7   Kilby, Noyce, and the Microchip

The cost of transistors gradually came down as production grew in the 1950s. Engineers began to realize, however, that transistors would soon run into a barrier. Buyers, especially in the armed forces, needed electronic devices to contain larger numbers of circuits in order to meet rising performance needs. These needs required reducing the size of the circuits. Circuits were wired together by hand, though, and therefore could not go below a certain size. The difficulty of packing more and more circuits to fit inside confined spaces came to be known as the "tyranny of numbers" problem, and engineers could foresee a plateau where electronics would remain if there was no way to overcome the barrier to miniaturization.

In the late 1950s, to break this barrier, the American military launched two research programs that relied on the leading electronics firms in the United States. The results were spectacular failures: in one, engineers pursued a circuitous notion of miniaturization that did not solve the problem, and in the other they tried to use materials in a way that was of scientific interest but proved impractical to engineer. The solution to the "tyranny of numbers" problem came instead from two engineers, each working for small firms that were more recent entrants to the electronics business. In the summer of 1958, Jack Kilby of Texas Instruments had the insight that transistors and other circuit elements could be made of a single material, such as silicon, instead of the different elements that the industry took for granted to be necessary. Robert Noyce of Fairchild Semiconductor in California recognized in early 1959 that a new process his firm had invented to manufacture transistors could allow whole circuits to be assembled by machine, eliminating the need for hand wiring. Combining the insights of Kilby and Noyce led to the *integrated circuit*, or *microchip*, a circuit on a flat chip that machines could make smaller than human hands could assemble.

The need of the U.S. Air Force for smaller guidance systems on missiles, and the need of the NASA space agency for a compact onboard computer to guide spacecraft to the Moon, provided the early market that the microchip needed to improve and come down in price. To give the integrated circuit a private civilian market, Kilby helped invent the first handheld consumer device to use a microchip, the pocket calculator, in 1967. In the following year, Noyce launched a new company, Intel, that became the leading maker of integrated circuits. He soon became the leading voice of aerospace and electronics firms in California, firms that pointed to a new kind of economy.

Jack Kilby: From Radio to Transistor

Jack St. Clair Kilby (1923–2005) was born in Jefferson City, Missouri, where his father, Hubert Kilby, was an electrical engineer. His mother, Melvina, worked as a hospital dietitian before starting a family. Hubert Kilby accepted a managerial position with the Kansas Power Company in 1927 and moved the family to Salina, Kansas; four years later, the Kilbys relocated to Great Bend in the center of the state. Like the Wright brothers, Jack Kilby and his sister Jane grew up in a house filled with books. As Jack began rising to his adult height of 6 feet 6 inches, he played football and basketball, and took up photography, which became a lifelong pursuit. Kilby's interest in electronics began in a massive snowstorm on April 7–8, 1937, which took down electric power lines. His father was able to communicate with utility crews only with the help of amateur radio operators, whom Kilby met by tagging along through the snow. He soon studied radio principles, assembled his own radio set, passed an amateur licensing test, and received a call sign.

In 1941, Kilby graduated from Great Bend High School and with two friends rode to Florida, and then to Washington, DC, in a Ford Model A car that Kilby had modified to add power to the engine (figure 7.1). The brakes failed on the way back to Kansas, requiring Kilby and his friends to slow the car by downshifting and using their feet. A more serious setback came when the future pioneer of the microchip failed the examinations that the Massachusetts Institute of Technology required for undergraduate admission. Fortunately, Kilby was able to enroll at the last minute in the electrical engineering program at the University of Illinois at Urbana-Champaign that autumn.

To build his electrical knowledge, Kilby remained in school after the entry of the United States into World War II on December 7, 1941. He joined the Army Reserve in December 1942 and six months later reported for active duty and entered the Office of Strategic Services (OSS), the wartime U.S. agency devoted to espionage and covert operations. After further training, Kilby went to Burma, where he served in a remote jungle outpost as a radio operator in support of Allied efforts to open a supply road to China. The opening of the road in 1945 found Kilby in China at war's end. Coming home, he graduated from Illinois in 1947 with a bachelor's degree in electrical engineering

Figure 7.1
Jack Kilby in his senior year of high school with his rebuilt Model A car. Courtesy of Ann and
Janet Kilby.

and joined Centralab, the research division of the Globe-Union Company in Milwau-
kee, Wisconsin. Globe-Union made hearing aids and electronics for radio and the new
medium of television. In his free time, Kilby completed a master's degree in electrical
engineering at the University of Wisconsin in Madison in 1950.[1]

Hearing aids needed very small circuitry, and Centralab was at the forefront of efforts
to make vacuum-tube circuits more compact. The lab devised a way to "silk screen"
connections by painting silver leads on a ceramic base in place of wires. But the lab
could not make vacuum tubes much smaller. When Bell Laboratories invited Centralab
and other firms to send engineers to a symposium in 1952 to learn about transistors,
Kilby attended and afterward shifted his focus at work from tubes to transistors. He
soon learned how to purify silicon and germanium and became proficient at making
transistors as well as designing circuits for them.

In 1958, though, Globe-Union warned its employees of an impending recession and the likelihood of cutbacks in work. By then, Kilby also wanted to go more deeply into work on transistor circuits, and Centralab was too limited in its focus. The leading computer manufacturer, International Business Machines (IBM), declined to offer him a position, but Motorola and Texas Instruments offered him jobs. Texas Instruments, the smaller of the two, seemed to provide more freedom, so in May 1958 Kilby moved with his wife and two daughters to Dallas, Texas, to join the firm.[2]

Robert Noyce: Taking Plunges

Robert Norton Noyce (1927–1990) was born in Burlington, Iowa, the third of four sons of the Rev. Ralph Noyce, a Congregationalist minister, and Harriet Norton Noyce, a 1921 graduate of Oberlin College. Ralph Noyce was pastor to a succession of Iowa churches until 1940, when he took a supervisory position in Grinnell, Iowa, with the regional conference of Congregational churches. Like Kilby, Robert Noyce took an early interest in radio, and at age 12 he also built a glider, similar in design to the Wright Flyer, with his older brother Gaylord, that they tried to fly. In high school, Noyce excelled at mathematics and science and graduated at the head of his class. He followed his older brothers to Grinnell College in their hometown (figure 7.2).

Robert Noyce was a star student at the college, excelling in science and finding time for extracurricular activities ranging from choral music to competitive diving. During the spring of his third year, his dormitory mates decided to hold a Hawaiian luau, and Noyce took a pig from a nearby farm that his housemates roasted. The next day, Noyce offered to pay the farmer. Theft of livestock was a serious matter in rural Iowa, though, and the farmer was angry. Grinnell's president interceded, as did Noyce's physics professor, Grant Gale, and as a result Noyce did not face criminal charges, but he was expelled for the fall semester of his senior year, which he spent as a clerk with an insurance company in New York. On his return to Grinnell, he completed a double major in mathematics and physics and graduated in 1949.[3]

Soon after Bell Labs made the invention public, Noyce learned of the transistor from Grant Gale, who learned of it directly from John Bardeen, a classmate of Gale's at the University of Wisconsin. Noyce was fascinated by the device and decided to apply to the Massachusetts Institute of Technology to study solid-state physics. Unlike Kilby, Noyce secured entrance, although he needed a year fully to catch up to the other graduate students. Noyce took a keen interest in the natural phenomena at work in electronics but wanted to put such knowledge to practical use. After finishing a doctorate in physics in 1953, he took a job with Philco, an electronics maker in Philadelphia,

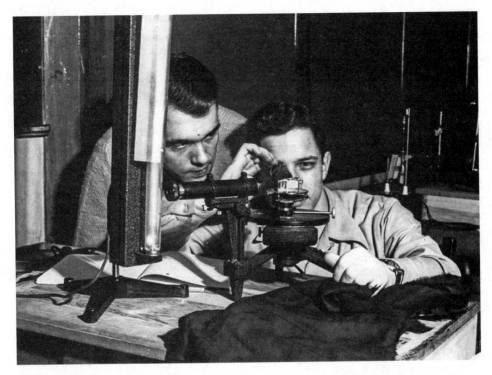

Figure 7.2
Robert Noyce (*right*) as a student at Grinnell College. Courtesy of Grinnell College Library and Professor William Case.

where he improved a new transistor that the company had designed. Over the next two years, though, Philco proved slow to manufacture it, and business difficulties curtailed new research.[4]

Noyce's name had come to the attention of William Shockley, who left Bell Labs in 1955 to form a company in Mountain View, California, near Stanford University. Shockley wanted to design and produce transistors of his own and he telephoned Noyce on January 19, 1956, to offer him a position. Noyce described the call from one of the three inventors of the transistor as "picking up the phone and talking to God." Noyce accepted and moved his wife and two children out to California.[5]

Shockley recruited 20 young engineers and scientists to his firm, all but two of whom were under the age of 30. In addition to Noyce, there was Gordon Moore, a chemist from the Applied Physics Laboratory of Johns Hopkins University, and Jean Hoerni, a Swiss-born theoretical physicist with two doctoral degrees. Both would play important roles later in Noyce's career. The young men around Shockley were unusual in having not only advanced training in science and engineering but also the practical skills to produce semiconductor electronics components themselves. They understood what it meant to fabricate as well as to design.

However, Shockley proved as difficult a supervisor as he had been in his last years at Bell Labs. Friendly at first, and a brilliant teacher who could simplify complex problems, he was inconsistent in his focus. He also discouraged independent work and was harsh to those who disagreed with him. After receiving the Nobel Prize in Physics at the end of 1956, he became more aloof and began to see his workplace more as a venue for advancing his own research than as a business for producing useful devices. Shockley's treatment of subordinates and his insistence on devoting work to a complex diode that was nearly impossible to manufacture finally caused eight of his younger men, including Noyce, to leave in the autumn of 1957.

Before resigning, the younger men found an investor, Sherman Fairchild, to support them in a new venture. The "traitorous eight," as Shockley called them, founded Fairchild Semiconductor (figure 7.3). The firm became one of several companies owned by Fairchild, son of the first chairman of IBM and himself the inventor of a camera that made mapping from the air practical. Noyce became the director of research in the new firm under a general manager who reported to the Fairchild Camera and Instrument Corporation in Syosset, New York.[6]

The Micro-Module and Molecular Electronics

The late 1950s were a time of growing uncertainty in electronics. The transistor began to enter use as an amplifier in consumer devices, such as radios and hearing aids, and it was about to replace the vacuum tube as a switch in electronic computers. However, Jack Morton, who took charge of transistor development at Bell Labs in 1948, observed a decade later that electronics faced a "tyranny of numbers" in which the need to pack more and more components into the small spaces of working equipment imposed a physical limit to the number of separate components that could be packed and hand-wired together. Morton thought that the transistor had solved this problem by replacing vacuum tubes, but in fact a "tyranny of numbers" threatened the transistor with the same problem.[7] Circuits using transistors could be smaller than tube circuits but still had to be assembled by hand, and as a result circuit parts could not be made and interconnected below a certain size.

The barrier to miniaturization was particularly worrisome to the U.S. armed forces.[8] During the early 1950s, in a program known as Project Tinkertoy, the U.S. Navy had

Figure 7.3
The Fairchild Semiconductor founders (*left to right*): Gordon Moore, Sheldon Roberts, Eugene Kleiner, Robert Noyce, Victor Grinich, Julius Blank, Jean Hoerni, and Jay Last. Courtesy of Magnum Photos. Wayne Miller, photographer.

tried to standardize some of the circuitry in vacuum-tube assemblies to make them faster to manufacture in an emergency.[9] The U.S. Army took this idea a step further in March 1958, when the Signal Corps, the branch responsible for army communications, contracted with RCA, a leading electronics firm, to develop the "Micro-Module." This was to be a small stack of transistor circuits on wafers (figure 7.4). The army needed radios and portable computers that were rugged and easy to maintain, and Micro-Modules could be snapped into place in equipment and replaced easily in the field. With a budget of $25 million (more than $100 million in 2010), RCA reported steady progress over the next five years, but the Micro-Module did not overcome the barrier to miniaturization and was obsolete by the time it was ready to use in 1964.[10]

The U.S. Air Force took a more radical approach to the "tyranny of numbers" problem. A scientist at MIT, Arthur von Hippel, had proposed in 1956 that the path to better electronics was not to work with existing materials but to create new ones at the molecular level that would simplify making devices.[11] His call for a more fundamental

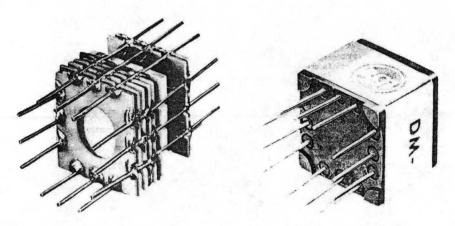

Figure 7.4
The U.S. Army Micro-Module, consisting of circuits on wafers stacked inside a small casing. Courtesy of U.S. Army.

approach encouraged the Air Force in 1959 to give another leading firm, Westinghouse, $7 million to create a way to achieve miniaturization that the firm called "molecular electronics." The aim of this program was to produce small "functional blocks" of new materials that would interface with each other directly and do away with conventional circuits interconnected by wires. Unfortunately, von Hippel's idea was too far ahead of its time. Over the next three years, the functional blocks proved impractical to manufacture and also had a limit to their possible miniaturization.[12]

Molecular electronics reflected the postwar idea that new science would lead to useful engineering. A more practical idea occurred to Geoffrey Dummer, an engineer at the Royal Radar Establishment in England. In 1952 Dummer proposed doping different parts of a semiconductor block to perform different circuit functions. He was unable to turn this idea into a working device. But Jack Kilby later credited Dummer with having anticipated the general idea of a solid circuit six years before Kilby himself turned his own attention to the problem of how to miniaturize electronics.[13]

The Monolithic Idea

Texas Instruments had begun in the 1930s as a maker of oil prospecting equipment under the name Geophysical Service. The company shifted its main focus after 1945. Patrick Haggerty, in charge of wartime aviation electronics for the navy, joined the new firm to help it pursue contracts for military electronics. In 1951, the company changed its name to Texas Instruments (TI), and a year later Haggerty secured a license from the

Figure 7.5
Jack Kilby in 1958, shortly after joining Texas Instruments. Courtesy of Texas Instruments.

Bell System to make transistors. Haggerty persuaded Gordon Teal of Bell Labs to join TI in 1953. Teal had developed a way to refine germanium that improved its use in transistors, and at TI in 1954 he produced a transistor made of silicon, a material harder to manufacture but more reliable than germanium.[14]

When Kilby arrived at Texas Instruments in May 1958 (figure 7.5), he discovered that RCA had recruited the firm to assist in the Micro-Module program and that he would be working on the firm's portion of the project. Privately, Kilby opposed the Micro-Module because it did not solve the problem of how to make circuits smaller.

He soon had an opportunity to think more deeply about an alternative. As a new employee, he had not earned vacation time, and when most of the firm went away in July, Kilby found himself almost alone for two weeks. "I was left with my thoughts and imagination," he later wrote.[15] To avoid working on the Micro-Module when everyone came back, and to solve the deeper problem, he began to rethink basic circuit design.

During the 1950s, much of the electronics industry worked on military contracts in which the goal was to produce the best equipment, not the cheapest. Having spent a decade working for a consumer electronics firm, where the cost of making things was all-important, Kilby came to Texas Instruments with an interest in finding solutions that were economical as well as efficient.[16] Kilby's new employer was a leading producer of semiconductor components, and he knew that the firm would benefit from an approach that used semiconductors in a more economical way.

Texas Instruments had innovated the silicon transistor and possessed as well the ability to make transistors from germanium. Kilby knew that other circuit parts, such as resistors and capacitors, could also be made from silicon or germanium, even though the two materials were not the best to use for these other components. For this reason, makers of electronic equipment used other materials. The more he pondered this fact, the more Kilby realized that the interconnection problem only existed because of the need to connect parts made of different materials. Making circuit parts out of the same material would eliminate this need. On July 24, 1958, Kilby wrote in his laboratory notebook: "Extreme miniaturization of many electric circuits could be achieved by making resistors, capacitors, and transistors & diodes on a single slice of silicon."[17] Kilby sketched how to form different circuit components on a chip by doping different parts to be n-type or p-type. He called his insight to make circuit parts out of a common substrate the "monolithic idea" (box 7.1).

When the firm returned from vacation, Kilby went to his supervisor, Willis Adcock, who approved an experiment. Using silicon transistors, and making silicon resistors and capacitors himself, Kilby demonstrated that a circuit wired together from these parts could work. The next step was to make a circuit on a single piece of semiconductor material, with different regions treated to function as different circuit parts. Kilby made a circuit on a single germanium slice because at the time the company did not have the silicon he needed (figure 7.6).

On September 12, 1958, before senior managers of Texas Instruments, Kilby turned on the device and used his circuit to turn a direct current into an alternating one. A week later, he made a circuit that could work as an on-off switch. Over the fall and winter, he was able to make these circuits in silicon.[18] In December 1958, after initiating what would be the second great electronics breakthrough of the mid-twentieth century

Box 7.1

The Monolithic Idea

By the late 1950s, electronics faced a limit to how small circuits could be made through hand-wiring. In 1958, shortly after he began work for Texas Instruments, Jack Kilby decided to see if he could overcome this barrier to miniaturization.

Until then, makers of electronic circuitry used germanium or silicon to make transistors. Different materials were found to be better for making resistors, capacitors, and other circuit parts. But these parts could be made out of a semiconductor as well. Kilby realized that if an entire circuit could be made from a single semiconductor material, a key barrier to miniaturization would be overcome. He called this insight the "monolithic idea."

Kilby formed a device on a single slice of germanium (below) that converted a voltage to a sine wave on an oscilloscope (i.e., converted a direct current to an alternating current), using a single transistor (Q_1), a capacitor (C_1), and three resistors (R_1, R_2, R_3). The resistors worked without doping, and for the other components of the circuit Kilby doped the semiconductor to create different kinds of junctions between regions treated to be p-type or n-type material.

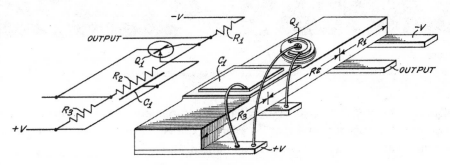

Kilby monolithic circuit diagram (*left*) and drawing (*right*).

In using germanium and silicon, Kilby used known materials to make his device, and he required no new science. His illustration used wires to interconnect the different parts, though, and as a result his patent did not show a fully integrated circuit, although he proposed in the patent that the wires could be replaced by metallic leads.

Source: J. S. Kilby, U.S. Patent No. 3,138,743 (1964).

Figure 7.6
Kilby experimental monolithic circuit. Courtesy of Texas Instruments.

(after the transistor), Kilby received a letter from Texas Instruments notifying him that his work was satisfactory and that his salary would be raised from $920 to $1,010 a month.[19] A modest man, he was no doubt pleased.

From Planar Process to Integrated Circuit

Jack Kilby demonstrated that the different components of an electronic circuit could be formed on a single chip of semiconductor material. However, under pressure to act quickly, Kilby filed a patent claim to the idea on February 6, 1959, with examples that still used wires, although he noted in his claim that metallic lines or "leads" could be laid down to replace them.[20] The contribution of Robert Noyce was to propose how to manufacture a circuit on a single semiconductor chip without wires. Together, the work of Kilby and Noyce removed the barrier to miniaturization.

Noyce and the other seven who launched Fairchild Semiconductor in late 1957 survived their first year in business by winning a contract from IBM to make silicon transistors for a new airborne computer needed by the U.S. Air Force. The transistors had to meet exceptional performance standards that other firms thought they could not achieve. Noyce confidently pledged to do so. The Fairchild firm divided into two teams to see which one could succeed first: Gordon Moore worked on producing a junction transistor with a p-type base, and the other led by Jean Hoerni worked on one with an n-type base. Moore's team was able to make its transistor first and Fairchild delivered the required 100 transistors by the summer of 1958.[21]

In meeting its contract, however, the Fairchild founders encountered a problem affecting all transistor makers at the time: only a small number of working transistors emerged from production runs. Tiny particles of dust and other contaminants ruined many of them. After his team lost the race for a new transistor, Jean Hoerni thought about this problem. In 1955, Bell Labs had discovered a way to make transistors by first laying down a thin mask of silicon dioxide over the semiconductor. Doping elements could be diffused through openings in the mask to create n-type and p-type regions underneath, with the mask protecting the material from heat applied afterward.[22] However, transistor makers regarded the silicon dioxide as a potential contaminant and etched it away when it was no longer needed. It occurred to Hoerni that this film might protect and improve the transistor if left in place. Tests soon showed that transistors were more reliable with the silicon dioxide layer left on them. In addition to leaving the mask, Hoerni designed the transistor to have the emitter, base, and collector on a single plane surface with the oxide layer, creating a device more practical to fabricate by machine in a manner similar to printing. Fairchild soon began making transistors in this way, in what Hoerni named the *planar process* (box 7.2).[23]

The firm's attorney asked Robert Noyce if the planar process might be useful for more than just making transistors. Noyce then realized that the process could be the key to a much larger breakthrough if it could be used to make an entire circuit on a single plane of semiconductor material under the oxide layer. On January 23, 1959, he wrote in his laboratory notebook: "In many applications now it would be desirable to make multiple devices on a single piece of silicon in order to be able to make interconnections between devices as part of the manufacturing process, and thus reduce size, weight, etc. as well as cost per active element."[24] Noyce outlined how aluminum leads laid down on a layer of silicon dioxide would not interfere with the silicon below and could interconnect the circuit parts on the chip through the leads and small openings in the layer. Using the planar process, machines could then make these circuits on chips of silicon at scales smaller than would be possible to assemble by hand. A team at Fairchild led by Jay Last overcame the engineering challenges and assembled a working version in 1961 (figure 7.7).[25]

Box 7.2

Planar Process and Integrated Circuit

In 1957, Robert Noyce and seven other engineers formed their own company, Fairchild Semiconductor, in Palo Alto, California, to make electronic components.

To make transistors, Fairchild engineers laid down and treated three layers of semiconductor material. The first layer was n-type, the next p-type, and the last n-type again. Metal contacts connected the emitter, base, and collector currents.

A serious drawback, though, was the large number of defective units in each batch of transistors. One of Noyce's colleagues, Jean Hoerni, examined the making of transistors more closely. Doping the semiconductor took place through a mask of silicon dioxide (a thin film that also formed naturally on the silicon from interaction with oxygen). Afterward, this mask was etched away. But the metal contacts and junctions then lay exposed to dust and other contaminants.

Over the summer and fall of 1958, Hoerni realized that the oxide layer could protect the contacts and the underlying transistor if left in place. Hoerni also moved the collector contact from the bottom to the top surface of the device. He called the result a *planar transistor* and he devised a method for manufacturing it. The result was a dramatic increase in the percentage of working transistors.

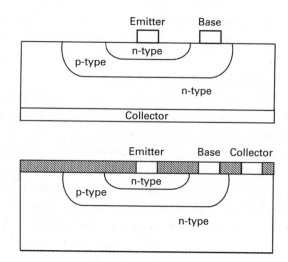

Transistor with silicon dioxide etched away (*top*) and left intact (*bottom, cross-hatched area*), with collector moved to the top surface.

Having the contacts all on one side also made it practical to make the transistors by machine in a method like printing. Robert Noyce then realized that if all of the components of a circuit could be made out of the same semiconductor, an entire circuit could be shaped, and connected together by metallic leads, using the planar process. But he came to the idea of a monolithic circuit in early 1959, several months after Kilby. Their two firms contested each other's patents until 1966, when they agreed to a settlement.

Source: Michael Riordan, "The Silicon Dioxide Solution," *IEEE Spectrum* 44, no. 12 (December 2007): 51–56.

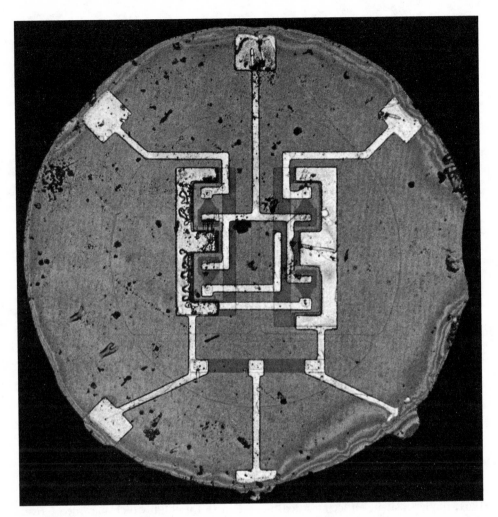

Figure 7.7
First integrated circuit produced by Fairchild Semiconductor, enlarged from its actual size of 0.125 inch in diameter (3.175 mm). Courtesy of ON Semiconductor, formerly Fairchild Semiconductor.

Kilby envisioned and demonstrated an entire circuit made of a single semiconductor. Noyce had a new method of interconnection, the planar process, with which he realized he could interconnect and manufacture an entire circuit on a chip. Combining the insights of Kilby and Noyce produced a new device, the *integrated circuit*, or *microchip*. Discoveries in quantum physics had suggested to the researchers at Bell Labs that a solid-state triode was possible. The transistor came from having an engineering objective, prior advances in semiconductor refining, and new scientific research. The integrated circuit combined an engineering insight, the use of a single material, with another engineering insight, the planar process. To innovate, Kilby and Noyce had to know the physics related to semiconductors, but they had to discover no new science in order to invent the microchip. The discrete transistor approached a barrier to its miniaturization after only a decade. The microchip removed the barrier and eventually became the basic component of electronics.

Like the transistor innovators, Kilby and Noyce had different personalities that influenced their approach to innovation. Kilby set aside time to read quietly. When he felt that he had perspective, he devoted himself intensely to solving a problem. Noyce was constantly in motion, often pursuing many ideas at once. He allowed insights to well up from his imagination, often in the middle of the night. However, Kilby spoke for both of them in observing: "A lot of solutions fail because they're solving the wrong problem."[26] From different starting points, Kilby and Noyce each identified a right problem and then contributed to its solution.

Texas Instruments came to recognize that the planar process was the best way to manufacture integrated circuits, and the firm licensed the process from Fairchild Semiconductor. But at first the two firms challenged each other's patent claims to the underlying ideas behind the microchip. Drafted quickly, the Kilby patent in February 1959 did not refer in as much detail to how to make the circuit, and in April 1961, the U.S. Patent Office awarded the patent to Fairchild, which had filed a rival claim on July 30, 1959, specifying the planar process. Texas Instruments appealed the decision, on the grounds that Kilby had conceived the monolithic idea earlier, and in February 1967 a higher body, the Board of Patent Interferences, ruled in favor of Kilby and the Texas firm. The U.S. Court of Customs and Patent Appeals then upheld the Fairchild claim, and the U.S. Supreme Court refused to hear the case in 1970. In 1966, however, the two companies agreed to cross-license their patents. By then, the microchip was too valuable for either company to depend on the uncertain outcome of litigation.[27]

Rocketing to Acceptance

The integrated circuit on a microchip attracted no customers at first. Private industry greeted the idea with many objections. The first integrated circuits were more expensive to produce than electronic circuits made of separate elements, and semiconductors were not the best materials for some of these parts. Nichrome made better resistors, capacitors were usually ceramic, and critics argued that each part should be made of the best material. Second, even with an oxide layer to protect them, batches of integrated circuits still had defects. Finally, integrated circuits threatened circuit designers with unemployment. "These arguments were difficult to counter," noted Kilby, "since they were basically true."[28] They were not relevant, though, for very long. Further work found ways to reduce the incidence of defective microchips to an acceptably low number, and engineers eventually found employment designing chips or devices that used them. Despite their relative inferiority, silicon resistors and capacitors also performed acceptably, and their drawbacks were soon offset by the advances in miniaturization and circuit density that microchips made possible.[29] But the cost of integrated circuits could not come down unless the market for them expanded, and it was the federal government that provided this initial market.

The navy expressed no interest at first, and the army initially thought the new chips might be a useful addition to the Micro-Module. However, the promised breakthroughs in molecular electronics never came, and the air force urgently needed a guidance computer that could fit into the nose cone of a missile. The U.S. Air Force gave Texas Instruments some initial funding to produce microchips, and in 1962 the firm received a contract to produce integrated circuits for the guidance system on the Minuteman II missile, part of a strategic triad of delivery systems for nuclear weapons (the nuclear missile submarine and the long-range bomber were the other two).[30]

A more spectacular boost came from NASA. The Apollo program to land astronauts on the Moon would not have been possible without a guidance and navigation computer small enough to fit into a small spaceship, and microchips helped to meet this need. Charles Stark Draper, director of the MIT Instrumentation Laboratory, ordered large numbers of integrated circuits from Fairchild in 1962 to fulfill the laboratory's contract to design the onboard Apollo computer (figure 7.8). The purchases by NASA enabled Fairchild to increase production of its microchips, bring down the price, and iron out early technical problems.[31] By the mid-1960s, the U.S. Army and U.S. Navy also finally embraced integrated circuits.

In 1967, Jack Kilby and a small group at Texas Instruments invented a handheld device that used a microchip, the pocket calculator (figure 7.9), and a consumer market

Figure 7.8

Apollo guidance and navigation computer, showing two display screens above the keyboard. A fold-down panel lists keyboard commands. Courtesy of NASA.

Figure 7.9
Pocket calculator invented by Kilby and his team in 1967. Courtesy of Texas Instruments.

for the chips soon grew.[32] Gordon Moore of Fairchild, who was initially skeptical of the prospects for the integrated circuit, soon recognized its potential for growth. He observed in 1965 that at the rate the integrated circuit had been miniaturized since 1961, a doubling in the number of transistors per unit of chip area could be expected approximately every two years for the foreseeable future.[33] This prediction, which became known as Moore's Law, held for the next four decades. Fairchild did not provide opportunity for promotion, though, and in 1968 Robert Noyce left with Gordon Moore to launch a new firm, Intel, that soon dominated the market for microchips and introduced a new kind of chip, the microprocessor, that made personal computing possible (see chapter 8).

A New Economy

The new microelectronics firms in Texas and California made a contrast to the industries established half a century earlier around the Great Lakes and in the American Northeast. The older industries employed large numbers of semi-skilled workers to mass-produce cars and other products. Although they also needed production workers, the new semiconductor and electronics firms in the Southwest and West depended on highly trained engineers and scientists to produce more sophisticated devices. A center of this new "high-tech" industry was the Santa Clara Valley around Stanford University on the peninsula south of San Francisco, California. Known informally by the 1970s as "Silicon Valley," and helped by a Mediterranean climate, the Santa Clara Valley saw rapid growth that attracted engineers and scientists. State support for the nearby University of California at Berkeley contributed to the growth of the wider region.[34]

A key figure in this growth was Frederick Terman (1900–1982) (figure 7.10), who joined the electrical engineering faculty at Stanford University in 1925. Terman became dean of engineering in 1945 and university provost from 1955 to 1965. From 1951 he arranged to make university land, faculty, and students available to firms that relocated to the Santa Clara Valley.[35] Terman believed that Stanford and other leading research universities were creating a new kind of economy by training and supporting the people who led new business firms in electronics and other rapidly growing fields. In 1965, the year of his retirement, he wrote:

> Such companies are characterized by the fact that a high fraction of the value of their output is the result of ingenuity, originality, and sophisticated engineering. As a result, an unusually large proportion of their employees are college graduates in engineering or science. Moreover, many of them have training at the graduate level. . . .
>
> These growth companies are the antithesis of the "mass production" companies. In the latter, most of the employees are concerned with production activities and perform simple repetitive tasks requiring only limited skills. The engineering costs then represent only a small fraction of the value of each article produced, and competition is based more upon manufacturing efficiency and marketing than upon creative contributions by the engineering staff.[36]

Terman saw new communities of highly educated people, with research universities at their center, as the future of American civilization. "Universities are . . . rapidly becoming more than mere places for learning," he wrote. "They are becoming major economic influences in the nation's industrial life, affecting the location of industry, population growth, and the character of communities."[37]

In fact, the new firms in Silicon Valley grew mainly from federal spending on national defense. The leading research universities also came to depend on federal support to conduct basic research in engineering and science and to subsidize tuition. Later, as they began to serve a civilian as well as military market, advanced technology firms also had to learn the mass-production and mass-marketing techniques that Terman associated with industries of the past. But Terman correctly foresaw that an advancing economy would need more people with higher levels of education and training.

Figure 7.10
Frederick Terman. Courtesy of Stanford University Special Collections and University Archives.

As head of the nation's leading microchip maker, Intel, Robert Noyce came to be known informally as the "Mayor of Silicon Valley," in which capacity he mentored a younger generation of computer entrepreneurs. By the 1980s, though, Japan and other Asian countries had begun to challenge America in electronics and other industries. Noyce warned that the newer products of "high technology" were as vulnerable as those of older industries to competing products manufactured in other countries.[38]

Jack Kilby did not seek the prominent role in business and public life that Robert Noyce achieved. After co-inventing the pocket calculator, Kilby found himself promoted to management positions that took him away from inventing. In 1970, he went on leave from Texas Instruments to devote himself to private study, in which he designed a solar collector for producing hydrogen as a source of energy. America's oil production (using conventional methods of extraction) had peaked in the 1960s, and in 1973 a brief disruption in oil imports underlined a new dependence of the United States on foreign sources of oil. Kilby's idea had promise but would have required a high initial cost to begin manufacturing, and he was unable to find backing. Kilby retired in 1983 and devoted himself mainly to his hobbies of photography and woodworking.[39]

Robert Noyce worked full tilt to the end of his life. In 1987 he helped found a consortium of U.S. semiconductor firms, Sematech, to increase industry cooperation in response to foreign competition. He died in 1990, just after sharing with Kilby the first Charles Stark Draper Prize of the National Academy of Engineering, awarded to the two of them for their contributions to the microchip.[40] Kilby went on to receive the Nobel Prize in Physics in 2000 for his part in the innovation. Before traveling to accept the award, Kilby stated that if Robert Noyce had lived, "we'd share this prize." In his Nobel Lecture, reflecting on the integrated circuit and his role at its inception, Kilby quoted the inventor of the laser, Charles Townes: "It's like the beaver told the rabbit as they stared at Hoover Dam. 'No, I didn't build it myself. But it's based on an idea of mine.'"[41]

8 The Computer and the Internet: The Age of Steve Jobs

With the innovation of the microchip, radios and other electronic devices using transistors for amplification became smaller and more efficient, and even more dramatic advances in a new field, electronic computing, became possible. In the early 1970s, circuits became small and powerful enough to enable the functions of a computer to be performed by a single microchip, the microprocessor, that brought computing within reach of individuals and not just large organizations.

The modern computer is an electronic machine that processes information very quickly in response to instructions. The first electronic computers in the 1940s used vacuum tubes and were employed as calculating machines. As transistors took the place of vacuum tubes, computers grew smaller in size and were able to carry out a wider range of instructions, but they still served the needs of organizations. During the mid-1970s, the founders of Apple Computer combined a microprocessor with a color television monitor to create a commercially successful personal computer. Competing personal computers from IBM then captured most of the market with software supplied by Microsoft and other firms. The networking of computers gave rise to the Internet.

By the early twenty-first century, computers and computer networking had begun to change how the modern world lived and worked. The industrial changes a century earlier had shifted a majority of Americans from rural to urban life and connected people more closely together through electric power, the telephone, radio, television, and the highway and personal car. Computing has not changed society as dramatically, but it has set in motion changes that may transform life in ways that are also cumulatively just as radical in the twenty-first century.

The Origins of Modern Computing

By serving as amplifiers, early vacuum tubes made it practical to communicate over long distances by telephone and radio. Vacuum tubes could also serve in switching circuits, to turn an electric current on or off. In 1919, the British physicist William Eccles and engineer Frank Jordan made a "flip-flop" circuit with two vacuum tubes, in which turning one on or off turned the other one off or on.[1] An electronic computer would be a series of switches, in which each switch could activate others to perform a task. The speed of electricity in these switches enabled the machine to bring a result much faster than computing manually (boxes 8.1 and 8.2).

Alan Turing, an English mathematician (figure 8.1), proposed in 1936 that a machine could compute any number if the problem could be expressed in the form of an *algorithm*, a finite series of unambiguous steps that followed preset rules. The idea proved its usefulness in World War II. Early in the war, teams of people had to make computations using mechanical desk calculators, work that was slow and tedious. Turing and other British researchers invented an electronic computing machine that could make computations much more quickly. For reasons of secrecy, however, the machine did not become public until the 1970s.[2]

Figure 8.1
Alan Turing in 1936. Courtesy of Princeton University Archives.

Box 8.1
The Computer: Basic Concepts

Algorithms

An electronic computer performs ("executes") a task defined in terms of an *algorithm*, which is a series of ordered steps that leads to a result. Each step in an algorithm must be unambiguous and the steps must finish in a finite amount of time.

An algorithm can define steps as a series of choices (e.g., if *x*, then do *y*; if not *x*, then do *z*). If set to add 2 + 2, an algorithm will follow a path of steps that lead to the result, 4. An algorithm may involve thousands of steps but the computation may take only a fraction of a second. A computer may execute many algorithms in succession.

The Computer

Physically, in executing a task, an electronic computer directs the flow of an electric current through switches, turning them on or off. At each switch, the electricity has a voltage (pressure) that is either "low" (off) or "high" (on).[a] Computers perform numerical calculations according to steps either entered as a programming algorithm or preset by the machine. The on or off states of switches represent the numbers to be computed and the procedures needed to compute them. Each step activates switches in the next step until the process ends in a result that the machine can display.

After 1946, early electronic computers had an "input" unit for entering tasks, a "memory" unit that stored usable functions, a central processing unit (CPU) with a "logic" unit and a "control" unit to perform the entered tasks, and an "output" unit to display the results. Modern computers have circuits that perform these basic functions.

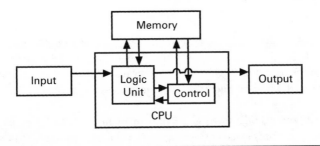

Source: M. D. Godfrey and D. F. Hendry, "The Computer as Von Neumann Planned It," *IEEE Annals of the History of Computing* 15, no. 1 (January 1993): 11–21. *Note:* (a) With a supply voltage of 5 volts, low on a switch can be up to 1.5 volts and high can be between 3.5 and 5 volts.

Box 8.2

The Computer: Logic Circuits

An electronic computer executes each step of an algorithm through logic *gates*. These consist of one or more switches through which current from an earlier set of switches can flow. Most are located in the CPU and the memory.

Each gate takes one or more entering currents and puts out a single current that is either "high" or "low" in voltage. The meanings of the input and output currents are assigned by the machine or by the program that is active.

There are three main logic gates used in computers: the "AND" gate, the "OR" gate, and the "NOT" gate. Nearly all algorithms can be carried out by a computer through arrangements of these three gates and four additional ones related to them.

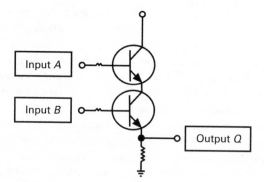

An AND gate with two transistor switches.

The output Q of the "AND" gate (*above*) is high when and only when the two input states, A and B, are both high. Otherwise the output Q is low. If a program specifies, for example, that the output of "high" at this gate means that the two inputs stand for equal numbers, the "AND" gate will allow the output to proceed if the two inputs represent equal numbers.

The OR gate outputs high when A or B or both are high; if none is high, the output is low. The NOT gate simply inverts a single input, from high to low or low to high. The other gates are not-AND (NAND); not-OR (NOR); exclusive-OR (XOR), which outputs high when either A or B but not both are high; and not-exclusive-OR (XNOR), which outputs low when either A or B but not both are high.

A flip-flop circuit (not shown) enables an entering current in one gate to reset the high or low state of another gate, which remains that way until reset. In this way, logic gates can serve as a computer's memory.

Figure 8.2
Inspecting tubes in the ENIAC computer. Courtesy of U.S. Army.

Across the Atlantic Ocean, John W. Mauchly, an assistant professor of electrical engineering at the University of Pennsylvania, received support from the U.S. Army in 1943 to build a computing machine to calculate trajectories for artillery shells. With a small group of assistants, Mauchly and a recent graduate student, John Presper Eckert, designed an electronic computer, the Electronic Numerical Integrator and Calculator (ENIAC), with more than 17,000 vacuum tubes and thousands of other components. The men on the team designed and assembled the machine; Betty Snyder Holberton and other women designed and implemented the initial switch settings to begin a calculation, a procedure called *programming* (figures 8.2 and 8.3).[3]

In the new machine, each calculation consisted of a finite number of steps, with triode vacuum tubes assigned to each step, where they acted as switches. Electric current entered the tubes of the first step, which activated more tubes in a chain reaction of steps that led to a final set of tubes whose on–off pattern gave the answer.

Figure 8.3
Programming the ENIAC computer. Courtesy of U.S. Army.

Calculations had to be set up by arranging cables on control boards, and more complex calculations required several days to prepare. But a simple addition then took 1/5000th of a second, and more complex computations could be done in less than a second. The ENIAC began operation on February 14, 1946, after the war had ended, but the machine proved that fast electronic computing was possible.

In early 1944, John von Neumann learned of the ENIAC project. An immigrant from Hungary, von Neumann was a distinguished mathematician at the Institute for Advanced Study in Princeton, New Jersey. He was working on the Manhattan Project at the time and needed a faster way to make computations. Von Neumann soon became an advisor to the ENIAC project, and he and the project engineers together conceived an improvement that became a vital part of later computers. This was the idea of a *stored program*, a pattern of on-or-off switch settings built into the computer, or set temporarily with reusable switches, that could automate more of its operation. Circuits that could hold a stored program came to be known as the computer's *memory*.

Later computers also embodied a 1945 proposal by von Neumann that the machines have five basic parts (box 8.1). An "input" device would translate instructions into a

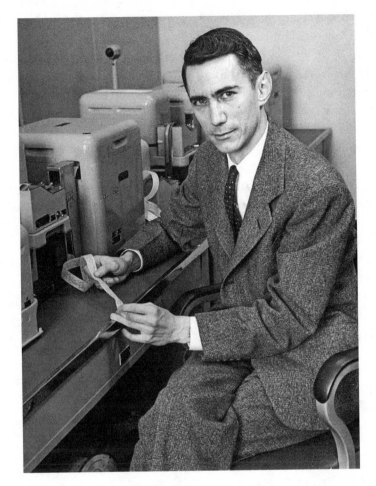

Figure 8.4
Claude Shannon in 1951. Courtesy of Getty Images/LIFE Picture Collection. Photographer: Alfred Eisenstaedt.

pattern of on-or-off switches, and a "logic" unit would perform calculations and other tasks in response. A "memory" unit would hold certain settings that could automate much of the computer's work, and a "control" unit would regulate the work of the logic and memory units. An "output" device would translate the results into some displayable form. Most later computers preserved these functions.[4]

In 1938 Claude Shannon (figure 8.4), a graduate student at the Massachusetts Institute of Technology, showed that switching circuits could be made to work best with a form of algebra, invented by the nineteenth-century English mathematician George

Boole, that used a binary (base 2) numbering system. Although the ENIAC machine worked with decimal numbers (base 10), it proved more efficient for later computers to execute tasks in binary choices and numbers. In 1948, now working for Bell Labs, Shannon found that it was possible to estimate a maximum number of binary signals that could be sent through a channel and then received without loss (box 8.3). The ability to estimate this amount was a crucial requirement for efficient transmission.[5]

The Mainframe Era

Mauchly and Eckert formed a company in 1947, soon acquired by Remington Rand, that produced the first electronic computer for a commercial market, the UNIVAC, in 1951.[6] During the 1950s and 1960s, a larger firm, International Business Machines (IBM), joined and then rose to lead the computer industry. Before the war, IBM had made and serviced mechanical office equipment. UNIVAC prompted IBM to bring to market its own vacuum-tube computers in the 1950s; and the IBM 650, leased to universities at a discount, greatly expanded the number of people who could program and use computers. By the early 1960s, computers using transistors began to replace those that used vacuum tubes. Although smaller than ENIAC, these machines could still fill a room and came to be known as "mainframe" computers.[7]

IBM's early computers and programs were not mutually compatible, though, and in 1964 the firm introduced System/360, a series of computers sharing common hardware and common programs or "software" (figure 8.5). Government and business firms soon leased or bought mainframe computers from IBM, and other computer makers soon produced competing machines with proprietary software of their own. Universities also bought mainframes for campus users to share. In 1965, the Digital Equipment Corporation introduced the PDP-8, the first commercially successful "minicomputer," a smaller machine that used transistors to provide some mainframe capacities at a lower cost. Minicomputers spread and added to their capabilities by employing integrated circuits.[8]

As computers improved, programming was hampered by the need to program them in the "machine code" of ones and zeroes that computers could understand. In 1951, Grace Murray Hopper (figure 8.6) at Remington Rand found a dramatic way to save programming time (box 8.4). Although not the first to reuse machine code, Hopper devised a program to reuse such code in collections that could be organized and retrieved. She also argued that these collections could themselves be organized in groups and retrieved, to make programming even more efficient. Then, after being advised that it wasn't practical to use words to instruct computers, she wrote a program

Box 8.3

The Information Revolution I

A New Kind of Information

Claude Shannon realized that solving numerical problems by machine would be simplest if problems could be reduced to algorithms in which each step was a binary choice. Electronic computing adopted this idea by assigning the numbers zero to the low state and one to the high state of a logic gate switch. Shannon called these two numbers binary digits, or *bits*.[a] He described these bits as units of *information*, giving the word the quantitative meaning that it now has in computing.

Binary signals also helped in another way. Amplification was necessary for electronic transmission but had the unwanted effect of also amplifying static or "noise," the tendency of a medium to interfere with communication, especially over long distances (see figure *below*).

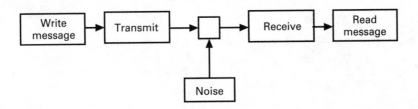

Shannon argued that discrete high and low states could be distinguished from the noise of interference more easily than continuous signals, making binary transmission more reliable for communication.

More Efficient Transmission

Messages also differed uniquely in their content. But Shannon went on to discover that the probability of particular numbers, letters, and other symbols occurring in any message could be reliably calculated.

Shannon's discovery made it possible to estimate a maximum amount of information (of any kind) that would fit within a given transmission channel capacity, and not undergo losses moving from sender to receiver. This made it possible to design computers and communication networks to move information efficiently.

Sources: C. E. Shannon, "A Mathematical Theory of Communication," *Bell System Technical Journal* 27 no. 3 (July 1948): 379–423, and 27, no. 4 (October 1948): 623–656. Flow chart is adapted from chart in *Bell System Technical Journal* 27, no. 3, 381. *Note:* (a) In early computing, a single number or letter consisted of eight bits or one *byte*.

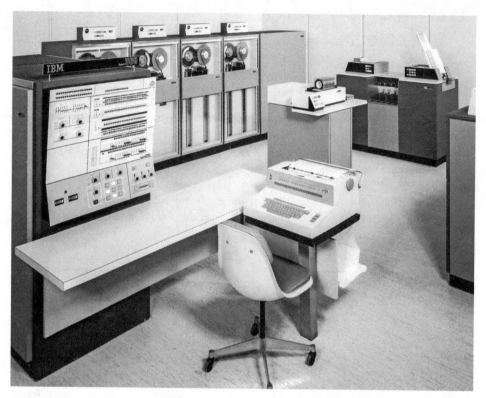

Figure 8.5
IBM System/360 (Model 40) mainframe computer. Courtesy of IBM Corporate Archives. © International Business Machines Corporation.

that could translate English words and sentences into the machine language of ones and zeroes.[9]

In 1957, John Backus of IBM introduced FORTRAN, a programming language for engineers and scientists, and in 1959, Grace Hopper helped produce a language to serve government and business needs, COBOL. Then in 1964 faculty at Dartmouth College wrote BASIC to give students an all-purpose programming language. By the 1970s, programs using languages such as these had evolved into two broad types of software: general *operating systems* for computers, and *applications* or programs designed to perform a narrower set of tasks within an operating system. Computer manufacturers usually provided the software to run on their own machines, but independent firms also sold applications and assistance in using them.[10]

Figure 8.6
Grace Murray Hopper in 1960 with a UNIVAC II computer. Courtesy of Unisys Corporation and the Hagley Museum and Library.

Box 8.4

The Information Revolution II

A New Kind of Library

Early computers had to be instructed in machine code, the ones and zeroes that the machines could understand. Reusing code saved time, but as tasks grew more complex, programming became more complex and time-consuming.

Grace Murray Hopper found a solution: to create "higher-level" programming, in which pieces of reusable code could belong to collections or "libraries" that could also be reused. In 1951 Hopper demonstrated a program, Arithmetic-Zero (A-0), that could organize and retrieve sequences of machine code by call number, as in a library.

Hopper also recognized the need for programming to use English words and sentences to stand for code, so as to make programming easier for those who found the earlier language of symbols and numbers challenging to use. Her Flow-Matic program proved that computers could use words and verbal sentence structures, as well as numbers, to call on libraries of reusable code and perform other operations. Flow-Matic was the basis for COBOL (Common Business Oriented Language), a widely used language for computer programming after 1960.

Software Development

Computer programs came to be known as "software" to distinguish the work of writing them more clearly from the design of the "hardware" in machines. The kind of linking program pioneered by Hopper's A-0 developed into the modern *compiler*, a kind of software program able to translate between high-level and low-level languages.[a]

The idea of placing reusable pieces of code into retrievable collections, which could themselves be grouped into higher level collections and retrieved, enabled programming to keep pace with the growing hardware capacities of computers and the need for increasingly capable software.

Sources: Grace Hopper, "Compiling Routines," *Computers and Automation* 2, no. 4 (May 1953): 1–5; and Kurt W. Beyer, *Grace Hopper and the Invention of the Information Age* (Cambridge, MA: MIT Press, 2009). *Note:* (a) Hopper called her A-0 program a compiler. Computer science now uses the term in a more general way to refer to a translator between levels of programming language.

Computers began to automate repetitive work, such as keeping records and making calculations, and made new services possible. As passenger jet air travel increased after 1958, the making of airline reservations and flight schedules in a timely way became more complex. In 1964, IBM produced SABRE, a computerized system for making and changing airline reservations. The new system borrowed technology from the SAGE system, a network of computers and tracking stations designed in the 1950s to give the U.S. Air Force warning of approaching enemy warplanes.[11]

Depending on its configuration, the IBM System/360 weighed from one to several tons, and other mainframes were similarly large. Mainframes and minicomputers needed specially trained personnel to operate them, and computers in the 1960s usually did their work in special rooms or computer centers. However, two advances began to break down this isolation in the 1970s: the postwar growth of television, and its joining to a new kind of microchip, the microprocessor, to create the personal computer.

Television

The most widespread new electronic device in the United States after World War II was not the computer but the television set, a box that received and displayed a moving image on a glass screen with accompanying sound. Electronic television originated in the work of two engineers, Philo Farnsworth and Vladimir Zworykin.

Born on a Utah farmstead, Philo Farnsworth (figure 8.7) taught himself electronics and conceived a television system, at the age of 14, in 1920. He patented the invention in 1930. Engineers knew that light that is reflected from an image onto a chemically treated surface could produce electric charges on the surface. Farnsworth devised a motion picture camera with such a surface inside that could capture the reflections of motion outside. The camera fed the charges into a circuit that could send them wirelessly to a receiver, which could project them as a flow of electrons to the inside surface of a receiving vacuum tube. Viewers could watch the moving images projected onto the receiving surface, which formed a glass screen when viewed from outside the receiver. A radio transmission could carry the accompanying sound.[12]

An immigrant Russian engineer, Vladimir Zworykin, conceived a similar system in 1923 and patented it but was unable to make a working model. In 1931 Zworykin became the director of television research for the Radio Corporation of America (RCA), where he invented a workable television system that could capture images in less light than the Farnsworth system.[13] RCA challenged Farnsworth's patent when he refused to sell it to the company, and a period of litigation followed that ended in Farnsworth's favor, requiring RCA to pay him a lump sum and licensing fees. At the New York World's Fair in 1939, RCA's president, David Sarnoff, demonstrated Zworykin's television, but World War II interrupted broadcasting until 1946. RCA engineers then improved the television camera again, and in 1947 Farnsworth's patent expired. In the late 1940s, television began two decades of rapid growth, and networks of broadcasting stations grew out of existing radio networks.[14]

Early television was monochrome (black and white) with a small screen in a large box (figure 8.8). In 1953, Zworykin invented color television, which became more

Figure 8.7
Philo Farnsworth circa 1920. Courtesy of Utah State Historical Society.

affordable in the 1960s. In color TV, the camera separated electrons into three flows to represent red, green, and blue colors and then recombined them in the receiver. By the 1970s, color cameras and receivers largely replaced monochrome ones, and television cameras and receivers began to use solid-state components in place of vacuum tubes.[15] In rural areas, a common antenna often supplied a number of households by cable, and the coming of satellite transmission after 1965 made individual reception possible from outer space. Cable and satellite television providers expanded into urban as well as rural areas in the 1980s and enlarged the number of channels available.[16]

Computers also became easier to use as a result of television. In 1951, a group led by Jay Forrester at MIT developed a computer as part of the SAGE air defense system that

Figure 8.8
The author (*right*) and his sister Elizabeth watching an early television. From a Billington family album.

employed video screens to display information. The use of video screens with computers spread in the 1960s.[17] In the 1970s, coupled with a new kind of microchip, the microprocessor, televisions (called monitors when used with computers) finally made it practical for computing to come to the general public.

The Microprocessor

In 1968, Robert Noyce and Gordon Moore left Fairchild Semiconductor to found a new company, Intel (shortened from Integrated Electronics), to produce microchips. The new firm focused on making integrated circuits for computer memories, and by 1970 advances in circuit design and manufacturing had enabled microchips to go from holding hundreds of transistors and other circuit components to holding thousands. However, Intel and other chip makers began to see a new engineering barrier on the horizon. Designers of electronic products typically created "chipsets" unique to each product and added new functions (abilities to perform new tasks) by adding more microchips. Each chip performed only the function for which it was designed, though, and as a result, devices began to contain more and more chips to perform more and more functions, many only infrequently used.[18]

A solution emerged at Intel from a contract to produce a chipset to go into a calculator designed by a small Japanese company, Busicom. The Japanese design had complex circuitry that Marcian "Ted" Hoff, the Intel engineer who received the assignment, believed to be impractical. It occurred to him that with reusable circuits on a single microchip, different kinds of software could carry out tasks that engineers were simply adding hardware circuits to perform. Hoff saw the Japanese contract as an opportunity to design a new microchip with this capacity. Two Intel engineers, Federico Faggin and Stanley Mazor, helped Hoff turn the idea into the 4004, a microchip with circuitry that could perform the functions of a central processing unit (box 8.5).[19]

The 4004 chip more than met the Japanese need for a calculator. Intel's leaders were hesitant to invest further in the idea, though, partly because they thought such a chip might have trouble competing with more powerful minicomputers, and partly because Intel could provide software support only to a small number of buyers. But in 1971, the company announced the new device, soon named the *microprocessor*. Over the next few years, Noyce and his colleagues realized that the new chip created an entirely new market, for small microcomputers. Outside developers would soon supply software for this market, relieving Intel of the need to do so. In 1972, Intel introduced the 8008 microprocessor; then in 1974, under the direction of a Japanese engineer from Busicom

Box 8.5
From Microprocessor to Personal Computer

The Microprocessor

By the late 1960s, engineers were adding integrated circuits (microchips) to increase the functions of electronic devices. In 1969, Ted Hoff, an engineer at Intel, realized that if a single microchip could perform different functions by running different software programs, additional microchips would not be needed to provide these functions. With two colleagues, Hoff designed a new microchip, the *microprocessor*, that could perform in this way and even serve as a central processing unit.

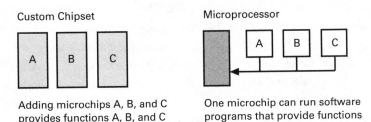

Custom Chipset

Adding microchips A, B, and C
provides functions A, B, and C

Microprocessor

One microchip can run software
programs that provide functions
A, B, and C

The microprocessor supplied the functionality of a larger computer on a single chip and made possible small "microcomputers" that the general public could afford.

The Personal Computer

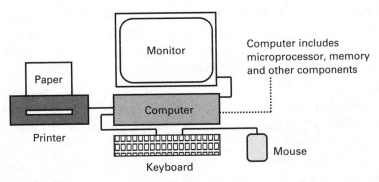

The personal computer evolved through different configurations. By the late 1980s (*above*), most desktop computers had a separate monitor and separate keyboard; a second input device, the mouse; and a printer that served as the output device.

who joined Intel, Masatoshi Shima, the company released the 8080 microprocessor, which became a lower-cost competitor to the minicomputer.[20]

As microchips grew more capable, however, circuit design on them became increasingly complex. This complexity was another looming barrier to miniaturization until 1979, when two engineers, Lynn Conway and Carver Mead, redefined the rules for microchip circuit design. Called VLSI (Very Large-Scale Integration), the new approach enabled the doubling of transistors every two years to continue. Conway's career nearly ended 10 years earlier, when she transitioned gender from male to female; a more supportive workplace at the Xerox Corporation's Palo Alto Research Center (PARC) in Palo Alto, California, enabled her work to continue.[21]

In 1975, MITS, a small firm in Albuquerque, New Mexico, began selling kits with which hobbyists could put together their own microcomputer, the Altair 8800, using the Intel 8080 microprocessor chip. The Altair could only display its results as a bank of lights on its front panel, but the kit made it possible for amateurs to assemble a small home computer (figure 8.9).[22] The momentous next step, taken by two hobbyists, Steve Wozniak and Steve Jobs, was to connect a microprocessor to a television and create a personal computer that did not have to be assembled from a kit.

Jobs, Wozniak, and the Apple Computer

Stephen G. Wozniak (1950–) grew up in Sunnyvale, California, part of the emerging "Silicon Valley" south of San Francisco. Still mostly fruit orchards, the area had begun to attract aerospace and electronics companies lured by Stanford University and a comfortable year-round climate. Encouraged by his father, an electrical engineer with the aerospace firm Lockheed, Wozniak learned basic electronic circuitry while still in elementary school. He became an amateur radio operator at age 11 and built an electronic calculator for a science fair two years later. After finishing high school in 1968, he attended two other colleges before enrolling at the University of California at Berkeley. In early 1973 Wozniak dropped out and found a job designing the circuitry of calculators for the nearby firm of Hewlett-Packard.[23]

In 1971, through a mutual friend, Wozniak met Steven P. Jobs (1955–2011), who lived in Mountain View, near Sunnyvale. Jobs's father was a mechanic and his mother worked as a bookkeeper. Steve Jobs finished high school in 1972 and two years later dropped out of Reed College in Oregon to work for the video game maker Atari, near Mountain View. Wozniak enjoyed the Atari game Pong, a version of table tennis played on a television set with a connecting control device. At the suggestion of Jobs, Wozniak designed an improved version of the game. In 1975, Jobs and Wozniak also attended

Figure 8.9
An MITS Altair 8800 microcomputer (with casing removed). Courtesy of Computer History Museum. Michael Holley, photographer.

meetings of a local hobbyist group, the Homebrew Computer Club, where the Altair 8800 was a focus of attention. Jobs saw the microprocessor as a path into business and he persuaded Wozniak to design a better computer than the Altair.[24]

Steve Wozniak had a long-standing fascination with computers and saw two engineering possibilities. First, the microprocessor brought within reach a computer that individuals could afford. Second, he believed that a television and a keyboard would provide a more intuitive display than the Altair's and would allow the user to do more with the computer. After determining that Hewlett-Packard had no interest in claiming the work as that of an employee, Wozniak designed the Apple I, a circuit board with a microprocessor and a small amount of memory. The buyer had to supply the keyboard and a television to display in monochrome. Jobs purchased components for the circuit board with money raised mainly from family and friends, and he negotiated with a local computer supply store to sell 50 of the boards over the spring and summer of 1976, charging $666.66 apiece, of which Jobs and Wozniak received $500.[25]

Wozniak soon improved the board by adding the ability to display color, and he began to design a more capable computer, the Apple II, that had a built-in keyboard. Jobs insisted that the device be enclosed in an attractive off-white plastic casing over which a monitor able to display color could rest (figure 8.10). However, the Apple II would require substantially more money to produce. Nolan Bushnell, the head of Atari, helped Jobs contact several potential investors until one, Mike Markkula, agreed to invest $91,000 and guarantee a loan for $250,000 more. A former marketing executive at Intel, Markkula incorporated the Apple Computer Company with Jobs, Wozniak, and himself as co-owners and the firm hired its first employees. Overcoming some hesitation, Wozniak left Hewlett-Packard to devote himself full-time to the new venture.[26]

The Apple II launched in April 1977 and was a phenomenal success despite its higher price of $1,298. Hobbyists took to the color graphics and used the machine to play video games. Wozniak and Jobs imagined a broader appeal, though, and in 1979 a small firm in Boston began to sell VisiCalc, a spreadsheet program for the Apple II that made the computer a business machine. A spreadsheet displays numbers in columns and rows and can perform calculations on any one number that are immediately reflected in the others, making financial projections easier. The Apple II (and improved versions of the machine) provided the new firm with most of its revenue over the next decade and enabled Jobs to raise millions of dollars from investors, including Arthur Rock, who had arranged financing for Fairchild Semiconductor and then Intel and who now agreed to chair the Apple board of directors.[27] In 1979, Wozniak survived the crash of a private airplane. He reduced his role in the company afterward and eventually retired in 1987, never having wanted a role in management.[28]

By 1980, Apple had several thousand employees, including many engineers. One of them, Jef Raskin, proposed a more radical successor to the Apple II that he called the Macintosh. The proposed machine would combine a monitor, a computer, and a fold-out keyboard in a single box.[29] Raskin urged Apple to learn from recent work at the Xerox Palo Alto Research Center (PARC). Xerox, a company known mainly for its photocopy machines, had started PARC in 1970 as an industrial research laboratory, and computer engineers at the lab had pioneered some key features of later personal computing. The most important of these was the graphical user interface.

Figure 8.10
Steve Wozniak in 1986, with an Apple II computer. Courtesy of Getty Images. Roger Ressmeyer, photographer.

Computers at the time required a user to enter instructions on a keyboard that were displayed on the monitor screen as a line of text. This became known as a "command line" interface, and errors had to be corrected by backspacing and retyping. Engineers at Xerox PARC designed software that instead displayed a simulated "desktop" on the screen, containing graphical objects that could be opened in "windows" by keystrokes or by moving a connected tracking device or "mouse" on a nearby table surface. Inside each window, the operator could perform tasks, such as composing text or drawing a picture, using the mouse or the keyboard. One of the engineers, Alan Kay, recruited nearby children to test "Alto" computers designed to operate in this way (figure 8.11).[30]

Figure 8.11
Children recruited to test an Alto computer at Xerox PARC. Courtesy of PARC, A Xerox Company.

Xerox made an investment in the Apple company in 1979 and as a result Steve Jobs was able to visit PARC, where he saw the Alto and its graphical interface. To his astonishment, and to the disappointment of the PARC engineers, Xerox management on the East Coast were slow to turn the computer into a commercial product. Jobs immediately gave a new Apple machine, the Lisa, a graphical interface, but at a price of $10,000 the Lisa did not find a market.[31] Jobs decided to make a less costly machine in the Macintosh. Raskin wanted the box to be portable and to sell for $1,000, and he left the project when demands from Jobs for more expensive features required a tabletop machine that would cost more. But under Jobs's direction, the project team designed a graphical interface for the Macintosh that made its operation efficient and appealing. The Macintosh launched in early 1984 with a separate keyboard and mouse but with a computer and monitor integrated in a single desktop box (figure 8.12).[32]

Figure 8.12
Steve Jobs in 1984, with Macintosh computers behind him. Photograph by Norman Seeff. Courtesy of Norman Seeff.

The Macintosh attracted buyers who liked its ease of use, especially as desktop publishing software for it became available, although its price of $2,500 was double the price of the original Apple II. In 1981, IBM priced its competing personal computer at $1,565, and the IBM machine (despite its command-line interface) soon proved more popular with the business market. By the mid-1980s, Apple's directors had a falling out with Steve Jobs, partly over the next new computer to produce, and Jobs left the company in 1985 to launch a new firm of his own, called NeXT. Apple settled into serving the smaller market it had created for elegant and easy-to-use machines.[33]

IBM tried to market a personal computer in 1975 for scientific use but sales were poor. The success of the Apple II prompted a new effort, and in 1981 a team led by Don Estridge came out with the IBM Personal Computer (PC). The PC used many nonproprietary parts in order to speed development, and its technical openness encouraged rival computer makers to find legal ways to reverse-engineer enough of what was proprietary to be able to manufacture "clones," or computers compatible with the IBM design. Although not an outcome the firm had planned, the IBM standard captured most of the business market for personal computers in the 1980s as a result.

Figure 8.13
Mark Dean in 1997. Courtesy of IBM Corporate Archives. © International Business Machines Corporation.

A key member of the PC team at IBM was Mark Dean (figure 8.13), an engineer who contributed the graphics capabilities to the original IBM PC. In 1982, Mark Dean became lead architect for the PC and added a component inside the next version, the PC/AT (for Advanced Technology), that allowed other firms to make external devices, such as printers, as well as internal parts, to work with the machine. Dean's innovation became known as the Industry Standard Architecture (ISA) system bus and helped propel adoption of the IBM standard.[34] Before it could advance, though, the original IBM Personal Computer and its successors needed an operating system, for which the firm turned to a start-up company founded by Bill Gates and Paul Allen.

Gates, Allen, and Microsoft

William H. Gates III (1955–) grew up in Seattle, Washington, where his father was an attorney and his mother was active in nonprofit work. As a high school student, the younger Gates shared an interest in computing with Paul G. Allen (1953–2018), whose father was a library administrator at the University of Washington. Gates and Allen learned to work with the BASIC programming language and consulted for the owners of a local computer. Allen attended the University of Washington but soon dropped out to work for Honeywell, a computer manufacturer in Boston, where Gates moved in 1973 to attend Harvard University.[35]

The Altair 8800 drew their attention when it came out in January 1975. The device did not have an operating system, and the Altair's maker, MITS, hired Paul Allen as a programmer after Allen demonstrated a simple operating system that he, Gates, and another Harvard student wrote for the machine. Gates then withdrew from Harvard and moved to Albuquerque to consult for MITS. He also managed an independent venture, Microsoft, that he and Allen founded later in 1975. Allen joined Microsoft full-time in 1976 and the new company grew by writing software for other computers.

The small firm moved back to the Seattle area in 1979 and opportunity soon struck. After failing to agree on terms with another software developer, IBM learned of Microsoft and asked the firm to write an operating system for the IBM Personal Computer. Paul Allen quickly purchased an operating system for the Intel 8080 processor, and with a small team to help them, Allen and Gates improved it for the IBM machine, which sold it as PC-DOS, short for Personal Computer Disk Operating System.

IBM soon altered the operating system for its own machines, but Microsoft retained copyright to its original system and was able to sell it, as MS-DOS, to the makers of IBM-compatible computers and earned substantial royalties as a result. Paul Allen survived cancer in 1982 and retired from the company, leaving Bill Gates to manage the firm as it entered a period of rapid growth (figure 8.14).[36]

Where Steve Jobs integrated his own computer hardware with software specially designed for it, Bill Gates made software that could work with computers made by different companies following the IBM hardware standard. Although most of its business evolved to serve IBM-compatible computers, Microsoft supplied software for Apple machines as well. From visits to Apple, Gates saw the graphical interface under development, and in 1985 Microsoft came out with a graphical interface integrated with its own operating system. The new product, called Windows, became popular by its third version in 1990, prompting a lawsuit from Apple for copyright infringement, but federal courts ruled against Apple's claim to key features of the interface.[37]

Microsoft went on in the 1990s to supply Windows to nine out of 10 personal computers in the world. This dominance was possible because personal computers and software for them were now affordable to millions of individual buyers as well as to business firms and governments. Microsoft produced its own applications to provide

Figure 8.14
Bill Gates at work. Courtesy of Getty Images. Doug Wilson, photographer.

word processing, spreadsheet, and database capabilities that were as important to the growth of desktop computing as the operating system and the computer hardware.[38] However, a new development, the Internet, soon challenged the desktop personal computer by providing a new medium of electronic communication.

The Rise of the Internet

The first American electronic computer, ENIAC, was a military project. The Internet also had its origins in military needs, as an initiative of the Advanced Research Projects Agency (ARPA), a small organization within the U.S. Department of Defense created in 1958 to explore unconventional ideas of potential military value. An office within ARPA funded computers and computing research at universities and private institutes. In its early years, the most important ARPA initiative in computing was an experimental network to link computers located in different parts of the country.[39]

During the 1960s, it became easier for more than one person at a time to use a computer, and shared access, or "time-sharing," became usual on university campuses.[40] ARPA recognized that if scientists and engineers in different parts of the country could share each other's computers over long distances, computing could be more widely available and more specialized in its use. The first director of computing research at ARPA, J. C. R. Licklider, envisioned a network of computers that could also serve as a storehouse of information.[41] ARPA faced two challenges, however, to achieving such a network. First, long-distance connections over the telephone system would be more expensive than local phone connections and could be prone to disruption. Second, the computers to be linked by the network were often different and incompatible.

The solution to the first problem was "distributed communication," an idea pioneered by Paul Baran, an engineer with the RAND Corporation, a civilian research institute in Santa Monica, California, on contract to the U.S. Air Force. Baran's idea responded to a military concern in the early 1960s, when fear of nuclear war was high and communication between key parts of the federal government and the armed forces needed to be reliable and secure. In traditional telephone communication, a continuous connection was necessary to complete a call or deliver a message. If the connection broke down, the call or message would be incomplete. To convey messages between computers, Baran called for breaking each message down into data blocks of uniform size. If each block knew its destination, it could bypass a broken link and take an available path to its destination. The different blocks could be reassembled at the end, in much the way that a message on paper, divided into separate sections, could be mailed in separate envelopes addressed to the same destination (figure 8.15).

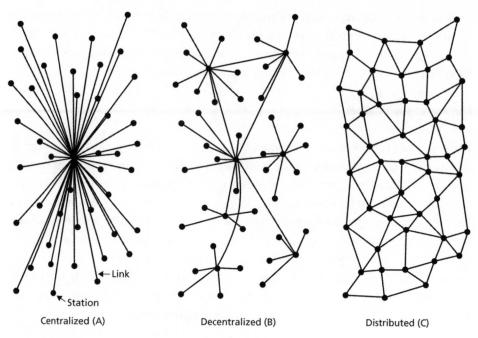

Centralized (A)   Decentralized (B)   Distributed (C)

Figure 8.15

Distributed communication. (a) Communication lines centralized around one point. (b) Telephone network of points linked through switching centers connected to each other. (c) A distributed communication network in which "packets" could follow any open path to a destination. Courtesy of The RAND Corporation.

Baran's idea provided a more secure way to communicate and also offered civilians in peacetime a more efficient way to share access to mainframe computers. Instead of each user needing a single uninterrupted connection to the computer, multiple users could share a more expensive high-speed line by breaking up their data and having the computer reassemble it on arrival, and then return data the same way. Donald Davies, a computer researcher in England, independently conceived of distributed communication in the mid-1960s and gave it the name *packet-switching*. ARPA learned about it from attending a conference where Davies was a speaker. Both Davies and ARPA then learned about Baran's earlier work.[42]

To reduce the problem of incompatible hardware, Wesley A. Clark, a computer scientist at the University of Washington in St. Louis, proposed that each computer center or "node" on the network acquire a standard minicomputer, called an Interface Message Processor (IMP), so that each node would connect to the network through the

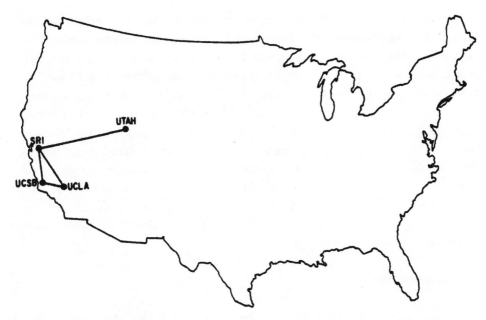

Figure 8.16
The first "nodes" of the Arpanet in December 1969: the University of Utah, the Stanford Research Institute (SRI), and the University of California campuses at Santa Barbara (UCSB) and Los Angeles (UCLA). Courtesy of DARPA and Bolt, Beranek and Newman.

same kind of machine. Each IMP could then link to the more varied computers at their location. Later, devices called *routers* provided this interfacing between local networks and long-distance ones. To communicate with each other, computers also employed a common software language, known as a *network protocol*.

In 1969, ARPA successfully tested an early version of its network, the "Arpanet," with a handful of nodes (figure 8.16). The agency added "Defense" to its name in 1972 and decided in that year that the network would need to transmit messages by radio and satellite as well as by telephone, and the protocol for telephone connections would not work with radio and satellites. In 1973–1974, Robert Kahn, a computer engineer at DARPA, wrote a program with Vinton Cerf, a computer scientist at the University of California at Los Angeles, who had helped to design the original protocol for telephone communication over the Arpanet. The new program came to be known as the Transmission Control Protocol/Internet Protocol (TCP/IP). The new protocols made communication between telephone, radio, and satellite networks possible. The larger resulting inter-network became known by the shortened term "Internet."[43]

The original purpose of the Internet, shared access to expensive mainframe computers over long distances, proved unnecessary as computers became smaller, more capable, and less costly for universities and laboratories to own. Scientists and engineers instead used the Internet mainly to communicate by electronic mail ("email"). In 1983, DARPA split its network, assigning part to the armed forces and merging the rest with a civilian network, the NSF-Net, managed by the National Science Foundation, which used high-speed cables leased from AT&T. After the breakup of the telephone monopoly in 1984, the successor phone companies began to offer data-carrying services over high-speed telephone lines, and cable television providers began to offer data services through their lines as well. In 1995, the NSF sold its network assets to private Internet service providers. There is no central control now, although an international group, the Internet Engineering Task Force, sets technical standards that are voluntary.[44]

The World Wide Web

The Internet still remained largely the preserve of universities and other research organizations until the late 1980s, when private companies began to offer consumers access to email and proprietary content over dial-up telephone connections from personal computers. In the 1990s, the Internet reached a much larger public.[45]

During the early 1960s, an electrical engineer at the Stanford Research Institute, Douglas Engelbart, invented the "mouse" tracking device, as part of a system consisting of a video screen on which a word or line of text could be clicked by the mouse to display another body of text on another page. A word or line that could link elsewhere in this way became known as *hypertext*. An outside network to make wider use of such linking did not exist, though, until the growth of the Internet in the 1980s.[46]

Timothy Berners-Lee (1955–) devised a new way to use hypertext. The son of two English mathematicians, Berners-Lee trained as a physicist but moved into computer programming. In 1980 he worked briefly at CERN, a physics laboratory sponsored by the European Council for Nuclear Research, near Geneva, Switzerland, and returned in 1984. Here Berners-Lee found hypertext a useful way to organize files on his local computer network, but he could not use it outside the network. TCP/IP made it possible to send documents from one computer to another, but to read the document still required having compatible software on each end. Email was more universal but lacked the ability to do more complex authoring. Instead of arguing for further standardization of hardware and software, Berners-Lee concluded that the Internet needed a simpler way to interconnect that would allow for the diversity that already existed.[47]

Figure 8.17
Timothy Berners-Lee, with early World Wide Web pages onscreen. Courtesy of CERN Press Office.

During the 1980s, a new kind of computer sharing had emerged. Instead of sharing the same computer in a room as in the 1960s, people now had their own desktop personal computers and used lines to connect to other computers, called *servers* when they held information that desktop computers could retrieve. To perform this retrieval, each desktop computer used software called a *client*.[48]

Berners-Lee designed a simple client program that he termed a *web browser* to retrieve text and other media from any computer server in the world that was open to the public over an Internet connection (figure 8.17). The web browser could read pages that had a few simple bracketed abbreviations, called tags, inserted in the text. These would be invisible to the viewer but would enable the browser to display the page and its content in a window onscreen. A page could display images and text together, along with other media, and certain tags could make a word, phrase, or image a hypertext link to other tagged pages anywhere on the Internet that was accessible. The browser needed to be compatible with its own computer, but the browser could retrieve tagged pages on servers that used different computers, operating systems, and authoring software. The servers holding these pages formed what Berners-Lee called the *World Wide Web*.[49]

Web servers, browsers, and pages achieved two goals. One was to enable engineers, scientists, and other researchers to share much more information between computers without having to share the same kinds of machines and software. The other was to allow easy hyperlinking across the Internet. With the help of Robert Cailliau, a Belgian programmer, and others at CERN, Berners-Lee refined the work and demonstrated it in the early 1990s.[50]

At the University of Illinois, the advantages of web pages were clear to two graduate students, Marc Andreessen and Eric Bina, who designed a simple web browser called Mosaic. Andreessen then created an improved version of Mosaic for commercial sale that he called Netscape Navigator. With the help of investors in California, Andreessen then went into business selling the Netscape browser to the public in 1994 (the browser later evolved into one called Mozilla Firefox). Public demand for Internet access grew dramatically as a result. In response, Apple launched a browser named Safari for use on its computers, and Microsoft launched a web browser, Internet Explorer, as part of its Windows operating system.[51]

During the 1990s, the number of web pages began to grow from several hundred to several million, as business firms began using the Web to advertise and market goods and services and as individuals began to put up personal "home" pages. After the year 2000, the cost of using high-speed lines began to come down, enabling telephone and cable television providers to offer higher-speed Internet access at a more affordable price. The use of dial-up access declined as the World Wide Web began to display pages that required higher-speed connections to view.

The most recent change in computing followed: the fusion of computing with mobile telephony. Steve Jobs came back to play a leading role in this change. Following his departure from Apple in 1985, Jobs produced an elegant personal computer, the NeXT, with all of the features that he desired. However, the devices were too expensive for more than a small market, although Tim Berners-Lee used one to make his first web pages and hold them as the first web server.[52] In 1997, facing the prospect of going out of business as its share of the market dwindled, Apple Computer invited Steve Jobs to retake control of the company. He quickly introduced a line of integrated desktop monitor/computers in brightly colored casings (a contrast to the universal beige) that attracted younger buyers and arrested the firm's slide. Apple also shifted its newer computers to a more stable operating system (based on Unix, a system developed at Bell Labs in the 1970s) and to microprocessors made by Intel.

In 2001 Jobs then launched a handheld player of digital music, the iPod, with an Internet storefront in 2003 to sell music inexpensively as electronic downloads. The music players became hugely popular. In 2007, Jobs introduced a new handheld device, the iPhone, that combined the features of an iPod and those of a mobile telephone. Mobile telephones had been invented in the 1970s but became widely used only in the 1990s when it became practical to make their circuitry more compact. The iPhone operated from a touchscreen and was the first "smartphone," or mobile phone with added capabilities, designed for a mass market rather than just for business use. Existing telephone companies with wireless data networks connected the new phones to the telephone network and the Internet. The iPhone, joined soon by competing handheld devices made by other firms, added cameras and other functions and created a global market of several billion people in the 2010s.

Although not himself an engineer, Steve Jobs had convinced his friend Steve Wozniak to design the first Apple computer, and Jobs went on to introduce the Macintosh with its graphical interface in 1984. Twenty-three years later, the iPhone popularized the handheld "smartphone," arguably the most ubiquitous technical innovation in the world of the early twenty-first century. By the time of his death from cancer in 2011, Jobs had established himself not only as a radical innovator but as an emblematic figure of the new digital economy and its growing influence in modern life.[53] At the end of his own life, he observed: "Sure it was great to make a profit, because that was what allowed you to make great products. But the products, not the profits, were the motivation." And he spoke for all innovators in adding: "Our task is to read things that are not yet on the page."[54]

The Computer as Innovation

Before the computer, industries tended to stabilize as they matured. In America, water control projects and the highway network were substantially complete in the 1980s, with their standards of design set by the 1950s. The basic designs of automobiles and airplanes were set by the 1930s and 1960s, respectively, although the design and performance of both continued to improve. The principles of electric power generation and power distribution were also established by the 1930s, and broadcast radio and television standards were set in the 1930s and 1940s, until digital standards replaced them early in the twenty-first century. In electronics, the vacuum tube reigned from the early 1900s to the 1950s, when transistors began to replace them in most uses. The integrated circuit or microchip spread in the 1960s and continues (at the time of this writing) to be the core component of modern electronics.

The computer was different. As the density of transistors on microchips doubled every two years from the mid-1960s to the end of the century, computing capacity increased spectacularly. In other branches of engineering, a doubling of capacity was a radical achievement. Yet in computing, a doubling of capacity every two years was almost an incremental innovation, although such doubling came at an increasing cost in the fabrication of microchips that made it possible. Microchips may be approaching a practical limit to their miniaturization in silicon. But research in fields such as quantum computing and other new technologies may continue to advance computing in dramatic (or cumulatively dramatic) ways, and the influence of computing in all areas of life is still unfolding and, in many ways, just beginning.[55]

Computing has had a controversial effect on society. Employment increasingly requires some ability to work with computing devices. Online business firms have expanded consumer choice and convenience but have also brought a degree of business concentration that resembles some of the older industries a century ago. New websites serve millions of people as means of social connection and as ways of finding information, but being on the Internet has also made personal privacy more tenuous and exposed business firms and governments to new forms of attack. Older industries a century ago also came with new benefits and dangers, and absorbing the change became a work in progress that continues with new technologies.

The spread of computer technology may give to the twenty-first century as distinctive a character as electricity and internal combustion gave to the twentieth century, and iron and steam to the nineteenth. In the developed world, and increasingly in the whole world, computing devices have become an inseparable part of everyday life. The full impact of this revolution has yet to be felt.

Conclusion

Modern engineering transformed American life in three broad periods. In the first, from the nation's independence until after the Civil War, engineers created new industries of iron and textiles, communication by telegraph, and new transport by steamboat and then railway. In the second period, from the 1870s to the 1930s, engineers changed America from a mostly rural and agrarian society to a mainly urban and industrial one, through new engineering led by telephony, steel and chemicals, motor vehicles and aircraft, radio, electric power distribution, and the use of new materials, steel and concrete, in building. The third period overlapped the second in the 1920s and 1930s, as engineers began to control great rivers for water and power and began to interconnect the nation with paved highways. Express highways and taller buildings came in the 1950s and 1960s. After 1945, nuclear energy introduced a new if controversial source of power, jet engines remade air travel, and rockets reached into outer space. A revolution in electronics and computing began that is still unfolding today.

Radical Innovation in Perspective

The great dams west of the Mississippi River made possible the phenomenal growth of the American West. But despite their immense size and power, these structures stand in a fragile relationship to nature, particularly in places where a changing climate could affect the amount of rainfall. Barriers to water may continue to be needed in other places, though, if rising sea levels and severe flood events cannot be managed in more natural ways. Some hydroelectric power will also be necessary as society relies more on sun and wind for energy, as these sources can vary in daily supply. Reservoirs of water form a steady reserve of power that dams can quickly tap to keep supply and demand for electricity in balance.

The Tennessee Valley Authority had a broader mission than any of the western dams: to relieve poverty across a vast region by using engineering to improve land use, supply electricity at a lower price, and control a major river. Not everyone benefited, but the agency did improve life for the many it served. Arthur Morgan's use of renewable energy and his emphasis on a more ethical way of life anticipated some twenty-first-century concerns. But Morgan's engineering depended on modern industry and his goal of local self-sufficiency did not come true. David Lilienthal saw the TVA as a means to prove that government could meet a vital social need that private enterprise had failed to provide. His efforts were a historic event in a debate that continues today over the role of government in society, but his model of regional authority did not spread to the rest of the country. After 1945, the TVA lost much of its original purpose. In recent years, the agency has tried to meet its obligation to supply electricity while contributing to more sustainable regional development.

The federal highway program, first with U.S. Highways and then with the Interstate Highway System, built a nationwide network to carry motor vehicles. Cities also rose into the sky with taller buildings. Daily personal mobility over longer distances became possible for millions of people and a mostly suburban civilization followed. Although Thomas MacDonald planned the highway network, its achievement was a partnership of his bureau with state engineers and the road-building work of large numbers of people, together with the rise of an automobile industry and fuels provided by fossil energy. Greater population density, new patterns of economic activity, and pressures on the natural environment could alter this way of life. However, new ways to continue a highly mobile civilization may not solve a deeper problem. As the author's father wrote in 1974: "Americans first moved west, then to the suburbs, and now just keep moving."[1] The future may require a better balance within modern engineering between the things that give freedom and mobility and the things that give a sense of belonging and place.

The nuclear power that Admiral Rickover designed for the U.S. Navy met a need that no other technology could provide. But Rickover recognized that a system to make use of nuclear energy would require extraordinarily rigorous standards of engineering and accountability to design and operate safely. Nuclear power did not meet the radical promise that it would make the energy too cheap to meter. The failure of a civilian nuclear reactor at Three Mile island in 1979 proved the need for Rickover's exacting concern for safety, as did later environmental damage from the extraction of fossil fuels in deep water and in new kinds of drilling on land. Although nuclear power plants in service do not emit greenhouse gases, they produce radioactive waste, and the long-term future of nuclear energy is uncertain. The lesson of Rickover's innovation was the need to be very careful with dangerous kinds of engineering and science.

The jet engine and the rocket owed their initial insights to the work of two individuals, Frank Whittle and Robert Goddard. The designs that flew airplanes and spaceships after 1945 were the work of a new aerospace industry funded initially for military purposes. Civilian airplanes propelled by jet engines largely replaced railways as means of long-distance transportation, and the Moon landing was a singular moment in the history of the world. The engineering of jet aircraft did not improve greatly after 1970, though, and the space program diminished afterward as well. Jet aviation seems likely to continue unless and until there are equally efficient and more comfortable alternatives, or modern life comes to require less speed. Ventures into outer space are set to continue at a slower pace, unless prompted by new security pressures or by an unusual event, such as the discovery of extraterrestrial life.

The technology that has defined the early twenty-first century is digital computing, which became more practical following the innovation of the transistor in the late 1940s. In 1945 the President's science advisor, Vannevar Bush, argued that basic scientific research, the study of nature for its own sake with no practical purpose in mind, was the source of all important technical innovation. The transistor has long been regarded as the star example of a technical advance preceded by pure scientific research (in quantum physics). But to imagine the transistor required a prior engineering image, the triode. New physics suggested in the 1930s that a solid-state triode might be possible, but wartime advances in semiconductor refining had to come first. John Bardeen and his colleague Walter Brattain had to investigate the natural properties of semiconductors more deeply, because the science in 1945 wasn't enough simply to apply. But in the development of the transistor, engineering and science were equally basic to radical innovation.

A decade after its invention, the making of transistors faced a barrier in how to miniaturize electronic circuits at scales smaller than human hands could assemble them. The American armed forces recruited leading electronics firms to overcome it, only to have them fail. Jack Kilby had the insight to see that simply making circuits on a slice of a single material would make it unnecessary to connect together parts made of different elements. Robert Noyce realized that a new process his company had just invented for making transistors could also make whole circuits at much smaller scales. Combining the two insights produced the integrated circuit, or microchip, and required no new science. Bell Labs chose to focus in the late 1950s on improving their landmark innovation, the transistor. Kilby and Noyce had the freedom and incentive afforded by new firms to innovate the microchip.

The transistor and the microchip illustrate how radical insight can occur in different settings. However, the different settings mattered to the engineering: the transistor would have been more difficult to develop in a small start-up firm, while the microchip did not require a large laboratory and an extended time for research. What the transistor and microchip had in common were the ability of their innovators to think radically, with the support of colleagues and higher management.

The general-purpose electronic computer began as a military project in the United States, and private industry began in the 1950s to manufacture machines. Computers remained the preserve of large organizations until the 1970s, when the microprocessor brought computing within reach of individuals. Steve Jobs and Steve Wozniak pioneered a commercially successful personal computer, and a small team of engineers at IBM then quickly produced a more open hardware design that evolved into the dominant standard for desktop computing, with the help of software produced by Bill Gates and others. In the 1960s, the Internet also began as a military project, and in the 1990s Tim Berners-Lee brought its connectivity to a much broader public through the World Wide Web. In the early 2000s, Steve Jobs integrated computing and telecommunication in handheld devices for consumers. Other manufacturers followed with devices of their own, and the devices spread to the world. Modern life is now increasingly digital.

America continues to depend on earlier engineering. Waterways and railroads still carry a large share of freight, and many of the industries that emerged a century ago remain important today. However, the period after 1920 differed in two ways. First, the federal government played an increasingly influential role in undertaking huge public engineering projects in the 1930s and 1940s. After 1945, the federal government funded the growth of new highways and new aerospace and electronics industries, and government continues to underpin advanced research in both engineering and science. Second, the trajectory of major innovation changed. Until about 1970, radical innovation had an outward and upward focus expressed by river engineering, long-distance highways, skyscrapers, jet airplanes, and space travel. Afterward, a more inward focus developed, typified by the design of electronic circuits at smaller and smaller scales, bringing with it a different kind of change. The role of government and the shift in focus pose important questions about the future.

The Role of Government

The federal government was relatively small before the 1930s, and until then private industry grew mainly on the strength of private demand for goods and services. This changed by the middle of the century. Spending on public works increased in the 1920s and 1930s, and after 1945 America required larger peacetime armed forces. National security concerns led in the 1950s and 1960s to federal funding of better highways, nuclear power, the aerospace industry, electronics, and an expansion of higher education. The size of the federal budget made it a central part of economic life.

During the late twentieth century, the role of government divided much of the public between two polarized views: that government is the stimulus to major innovation, or that it is a hindrance. Insofar as this dispute is about the actual record of what government did, the technologies covered in this book show that government played a role vital to all of them. But the role of government varied. In electronics, engineers in private firms innovated the transistor and the microchip, but to come down in price the new devices required federal purchases in bulk quantities. The computer and the Internet began as military programs but required private investment to evolve into new civilian industries. The federal government and private firms interacted in a way that is hard to reconcile with a polarized view of government.[2]

Innovation is now seen by many as the product of a managed process or "innovation system" that links government, universities, and industry in advanced research, with expert peer groups having a large influence.[3] Federal spending on research in engineering and science increased during the Cold War for reasons of national defense, and today it continues not only to support national security but also to develop new technologies to propel civilian economic growth.[4] There are two questions that the examples in this book raise about the future of federally supported research.

The first is whether science should be said to come before engineering. Scientific discoveries in nuclear physics directly inspired the search for nuclear power, and science played a crucial role in the transistor. But engineering design in both cases came from a different and independent kind of imagination and insight. A knowledge of scientific principles and natural properties was necessary to radical innovation in public works, new energy, aerospace, and electronics, but science did not provide the insight necessary to design. Engineering and science may become increasingly intertwined in the future, but the two should be seen as equally basic to fundamental research, rather than one being seen to derive from the other.

The second question is whether radical engineering insight is best served by institutional arrangements that seek to manage and sustain innovation as a kind of flow. Rare insights and/or one-time goals are characteristic of radical innovation and are not part of a flow in any normal sense. A system designed to manage innovation as a flow has a machine image in mind, with innovation seen as coming from a kind of assembly line. This model is better at producing more limited and frequent kinds of technical change. Even the federal "DARPA model" of giving researchers broad latitude typically asks for results in five years or less.[5] To continue, radical insight may require a place for a less process-oriented concept of innovation.

The Trajectory of Innovation

The other way in which the innovations in this book differed from those before them was in long-term direction. From the 1920s to the 1960s, engineering continued an earlier pursuit of innovations that reached outward, through public works such as dams and highways that remade the landscape, and upward through skyscrapers, jet airplanes, and space travel. In the late twentieth century, the technical emphasis in the United States and other advanced countries shifted inward, to innovation at smaller and smaller physical scales, the preeminent focus being circuits on microchips. A reduction in federal spending on defense and big infrastructure projects explains part of the shift, but the larger economy also changed, possibly reinforcing the new focus.

By the 1970s, industries that had begun in the early twentieth century reached maturity, in the sense that production no longer expanded as rapidly as it had in the years from 1900 to 1929 (and did again for a time after 1945). The leading sector in America, the auto industry, evolved from radical innovation in its early years to an underlying standardization and emphasis on incremental change in the post–World War II era, and motor vehicles per person roughly stabilized after 1970.[6] A change in employment also occurred. In the early twentieth century, people leaving farms could find semi-skilled work in the new steel, motor vehicle, and electrical goods industries. In contrast, the growth industries of the late twentieth century tended to employ fewer people domestically and required higher skills. The older industries also needed fewer people as a result of more stable demand and continuing automation.

Americans tend to think of technology in terms of machines, which are mass-produced to work everywhere in the same way, are usually owned and used privately, and become obsolete quickly. But if the challenges of the twenty-first century have to do with economic security and sustainability, there may be renewed interest in problems of place, in public needs, and in things that endure. These relate to the structural needs of society.

This book leaves out major advances in agriculture and medicine and touches only briefly on the military sphere. The idea of interchangeable or standardized parts, so central to production in the twentieth century, in fact began in the nineteenth-century manufacture of firearms.[7] Crucial to the improvement of modern life was a better understanding of plants, animals, and microbes, along with medical discoveries in the causes and treatment of illness and many disabling conditions.[8] But advances in public health also depended on the civil engineering of water and sanitation, and it was mechanization in farming and in the transportation of food that ended famine and freed majorities to work in cities and suburbs. Some of these advances now raise questions of concern today: about the sustainability of water and food supplies, about health security, and about emerging technologies that could be weaponized. These concerns cannot be more fully addressed in the present volume, but they deserve attention in the future.

Prospects

What does the future hold? Robert Gordon of Northwestern University has observed that many of the achievements of the industrial revolution were one-time events. Gordon argues that society will not need to reinvent the automobile; the future will only bring improved versions of it. Modern water systems and sanitation, roads and bridges, electricity, and other technologies that began more than a century ago may also be improved in how they work but in their essentials shouldn't need to be reinvented. Gordon argues that society will settle eventually into a pattern of activity characterized by more gradual technical change and slower growth.[9]

Although the trend of the last two centuries supports Gordon's thesis, it is important to underline the fact that the trend of modern engineering has also been to emancipate human beings from unwanted work. The industrial revolution substituted machines and inanimate energy for the muscle power and energy of humans and animals. After two centuries, almost all people in the developed world can now be supported in non-farm occupations. If computing raises the level of skill at which machines can reliably perform, then machines will increasingly replace human labor. At the same time, they could also relieve more people of unwanted work.[10]

The pattern in which the descendants of those displaced from older occupations find employment in new ones may continue. But the longer trend of modern engineering points to the achievement someday of a time when work itself becomes more of a choice than a necessity. If machines someday reduce the need for human employment to a fraction of its present level, in the way that machines in the twentieth century reduced the number of people needed in agriculture, then it seems likely that the benefits of such automation will have to be shared, and from that time onward work could become much more of a voluntary choice. This time may be a distant prospect, but it might be helpful if the prospect begins to reshape how society thinks about technology and the means and purposes of its advance. The idea of technical advance as the means to an always-moving frontier of needs and wants is only true in a relative sense: in fact, there are basic human needs and wants that may be possible to meet in a more permanent way, freeing people for other activities that give other kinds of meaning to human life.

Modern society will first have to bring itself into a sustainable balance with nature. The great innovations have had unanticipated consequences, and the greatest engineering challenge of the twenty-first century may be to relieve the pressure of modern civilization on the natural environment. This pressure is having a dangerous effect on climate and has made it easier for diseases to jump from animals to humans. New dangers may also appear as a result of technical advances that come before society is ethically prepared to manage them. This risk is apparent in the use of computing for surveillance and is also likely to arise in biological engineering, in the synthesis of living things and in the manipulation of the human genome. These prospects underline the responsibility of engineers and scientists who learn how these things can be done. Although constrained by the need for efficiency and economy, modern engineering is not the discovery of an inevitable "one best way," or the operation of a supposed technological imperative. Engineers have a degree of freedom and responsibility for the choices they make, as do whole societies.

A level of knowledge and skill higher than that required a century ago is now necessary to most technical work. But skills alone will not generate productive new insights. Currently taught engineering knowledge and skills impart the means to solve problems that are, in an underlying sense, already familiar. Such knowledge can be helpful only up to a point in tackling problems that go beyond what is known. Radical innovation requires a kind of insight that the study of actual engineers and their works may best explain and impart. Understanding how earlier innovators overcame barriers and challenges can give engineers, and a wider public, insight into how modern civilization has evolved and what major innovation involves.[11]

Human beings are naturally curious and there will continue to be challenges and opportunities that inspire unconventional insights. To study the history of radical engineering innovation is to learn how the deepest kinds of technical change can occur. The deepest challenge may be what to do with this change.

Acknowledgments

I owe first and foremost a debt of gratitude to my father, the senior David P. Billington, who served in what is now the Department of Civil and Environmental Engineering, in the School of Engineering and Applied Science, at Princeton University from 1958 to 2013. In his teaching and research, he developed the approach taken in this book and outlined much of its content. Regrettably, he could not co-author this last volume in the trilogy that he began on modern engineering, but his exploration of the topics and insistence on clarity and simplicity inform the work, and he read and approved early draft chapters.

The Hon. James A. S. Leach, as chairman of the National Endowment for the Humanities, made a Chairman's Grant to me for which I am immensely grateful. The grant helped make the research and writing of this book possible. I also owe a debt to the Walter L. Robb Senior Fellowship, for which my father was encouraged to apply by Dr. Norman L. Fortenberry, founding director of the Center for the Advancement of Scholarship on Engineering Education under the National Academy of Engineering. The Robb Fellowship enabled us to carry out early research for this book. Any views, findings, conclusions, or recommendations expressed in this book do not necessarily reflect those of the National Endowment for the Humanities, the National Academy of Engineering, or any other grant-making organizations.

Grants to my father from the National Science Foundation's Division on Undergraduate Education, and from the Alfred P. Sloan Foundation, supported the teaching and research that made the present book possible. Doron Weber at the Sloan Foundation deserves special thanks for a long commitment that this volume brings to a conclusion. I would also like to thank the Hon. John W. Kern III and the Hon. B. Paul Cotter for their invitations to me, beginning in 2007, to deliver an engineering lecture to the Harold R. Medina Seminar (from 2012 the Kern–Medina Seminar) for State and Federal Judges, held annually in June on the campus of Princeton University, which sharpened my ability to present engineering ideas to a distinguished audience.

This book draws on the course "Engineering in the Modern World," which my father began teaching at Princeton in 1985. Now taught by Professor Michael G. Littman in the autumn term, the course provides an overview of modern engineering innovations. During a leave of absence by my father (2007–8), I served as a preceptor (seminar leader) in the course, and Professor Littman's advice and encouragement have been vital to the completion of this book in the years since then. I would also like to thank Professor Maria Garlock, who teaches a spring term companion course, "Structures and the Urban Environment," for her assistance in my research for this book and for another book on the Golden Gate Bridge. I would also like to thank Professor Sigrid Adriaenssens and Professor Branko Glisic for their interest and support.

For their commitment to "Engineering in the Modern World" and its scholarship, I share my father's gratitude to the Department of Civil and Environmental Engineering, the School of Engineering and Applied Science, and the faculty Council on Science and Technology at Princeton University. I would also like to give special thanks to Professor Paul Prucnal for reviewing a chapter in this book. During the 1990s, President Harold Shapiro enabled the course and its associated research to grow, and Presidents Shirley Tilghman and Christopher Eisgruber sustained this teaching, as did Deans of Engineering and Applied Science James Wei, Maria Klawe, and H. Vince Poor, Associate Dean Roland Heck, and Professors Peter Jaffe, Michael Celia, and James Smith as department chairs. I also want to thank preceptor Thomas Roddenbery, laboratory director Joseph Vocaturo, and administrative assistant Kathy Posnett for their assistance. Timothy Manning in Media Services assisted the senior and junior Billingtons to give lectures, and J. Wayman Williams, Princeton Class of 1947, provided crucial photography and publication assistance. The graduate students who helped teach the course and the many undergraduates who took it proved the approach and enriched its content. The admiration and affection of so many students for my father after their graduation was a sustaining influence not only for him but also for me.

My father's classmates from the Princeton Class of 1950 and other Princeton alumni deserve special thanks for their encouragement and support of my father's work before and during the period in which I researched and wrote this book. From the Class of 1950, I would like in particular to thank Charles Rose, Jon Lovelace, Stuart Duncan, Peter Erdman, and David McAlpin, and from the Class of 1960, Preston Haskell. In the revised edition of his classic text, *Common Sense on Mutual Funds*, John C. Bogle, Class of 1951, honored us with very kind words about our 2006 book, *Power Speed and Form*. Gordon Y. S. Wu, Class of 1958, endowed my father's faculty chair from 1996 to 2010, and I owe a very special debt to Professor Norman Sollenberger for mentoring my father throughout his career and for sharing some of his experiences with me.

The Hill School in Pottstown, Pennsylvania, the New York State Regents External Degree Program (now Excelsior College), and the faculty of the Paul H. Nitze School of Advanced International Studies of the Johns Hopkins University, and the faculty at the University of Texas at Austin, prepared me for work as a historian, as did working with the staff of the Woodrow Wilson International Center for Scholars and the Joseph Henry Papers, both at the Smithsonian Institution in Washington, DC, before I enrolled in the doctoral program at the University of Texas. I am grateful especially to my doctoral dissertation advisor, Professor W. Roger Louis, at the University of Texas at Austin, for his guidance and support.

My journey in life would have been impossible without additional mentors. I owe a very special debt to Professors Cyril E. Black and Julian P. Boyd of Princeton University, Stephen C. Flanders of WCBS Radio in New York City, and Roswell B. Wing of the U.S. Department of Commerce. I would also like to thank Orest Pelech, S. Frederick Starr, R. Bayly Winder, and Jay Bleiman for their help at important moments in my life, along with Thomas G. Ruth of The Hill School.

I would like to give special thanks to Ann and Janet Kilby, daughters of Jack St. Clair Kilby, who graciously supplied and gave permission for me to include a photograph of their father. I am also grateful to Ed Millis, a friend and biographer of Kilby, for helping me to get in touch with the Kilby family. The distinguished photographer Norman Seeff gave permission for me to choose one of a set of iconic photographs that he took in 1984 of the late Steve Jobs. I would like to thank his business partner, Charles Hannah, who kindly supplied the photograph I chose and authorized its use. I would like especially to thank Dr. Mark E. Dean, who helped design the IBM Personal Computer, for information on his work.

Libraries were vital to the author's research and writing. The Princeton University Engineering Librarians Adriana Popescu and Willow Dressel were especially helpful, as were Associate Librarian for Rare Books and Special Collections Ben Primer, University Archivist Daniel Linke, and Curator of Manuscripts Don Skemer. I would also like to thank the Library of Congress Manuscripts Librarian James Hutson for assistance with the papers of Jack Kilby and with many other inquiries. The interlibrary loan service of the Santa Monica Public Library and the collections in the Charles E. Young Research Library at UCLA, the Los Angeles Public Library, and the Southern Regional Library Facility at UCLA also contributed to the research for this book. I am indebted to the scholars, librarians, and archivists identified by chapter below.

For the first chapter on river engineering, I owe a very special debt to Professor Donald C. Jackson of Lafayette College. His scholarship on the history of American dams, and his book, *Big Dams of the New Deal Era* (2006), co-authored with my father, served as the basis for the chapter. *Big Dams* began as a project jointly sponsored by the federal Bureau of Reclamation and the U.S. Army Corps of Engineers. My father and I both received valuable assistance from Brit Storey, historian of the Bureau, and Martin Reuss, historian of the Corps. I would also like to thank Virginia Berkeley for permission to use a photograph of Jack Savage taken by her father, and Thomas Midthun, Jr., for two photographs taken by his father. Patrick Jacobs of the Bureau of Reclamation also provided photographs for the chapter.

For the second chapter on Arthur Morgan and the Tennessee Valley Authority, I would like to thank Deborah Janning and Janet Bly of The Miami Conservancy District in Dayton, Ohio, for photographs; and Deborah Janning and Brenda Gibson for reviewing a section of the chapter. I am also grateful to Scott Sanders of the Antioch College Library for photographs of Arthur Morgan and the first TVA Directors. I could not have written the chapter without the archival research on the TVA performed by Professor Abbie Liel, now at the University of Colorado, when she was a graduate student at Princeton, with the generous support of Princeton Class of 1960 alumnus Preston Haskell. Susan Elder of TVA River Operations provided valuable engineering data on Wilson Dam. Patrick Fahy at the Franklin D. Roosevelt Library in Hyde Park, New York, along with Maureen Hill of the National Archives in Atlanta, supplied images, as did the Library of Congress and the TVA itself.

For the third chapter on highways, I would like to thank Andrea Henry and Rick Murray of the Iowa Department of Transportation, Richard Weingroff of the Federal Highway Administration, and Shubangi Kelakar of the California Department of Transportation Library for photographs. I would like especially to thank David Goodyear and the late Arvid Grant for a photograph of the East Huntington Bridge designed by Grant, and I am especially indebted to Yasmin Sabina Khan, daughter of engineer Fazlur Khan, for having written an excellent biography of her father. I am very grateful to William F. Baker, partner of Skidmore, Owings & Merrill in Chicago, and to Adri Jevtic at SOM, for assistance, and to Scott Lewis of *Engineering News-Record* for permission to reprint a cover of the magazine. I would like to thank J. Wayman Williams, Princeton Class of 1947, for a photograph of the John Hancock Center in Chicago and for his generous assistance with computing and graphics software.

For the fourth chapter on nuclear power, I would like to thank Justin H. S. Breaux of the Argonne National Laboratory; Robert Cressman, John Greco, and Matthew Staden of the U.S. Naval History and Heritage Command; Suzette Kettenhofen of the Navy Office of Information; and Nathaniel S. Patch of the National Archives in College Park, Maryland, for photographs or help with documentation on Admiral Hyman Rickover. I would like to thank Wendy Gulley and Mike Riegel of the Submarine Force Library and

Museum in Groton, Connecticut, for information on the USS *Nautilus* (SSN-571). I am also grateful to Michelle Mittrach and Alan Carr of the Los Alamos National Laboratory, Jacqueline Loop of the Idaho National Laboratory, and the U.S. Navy and the U.S. National Archives for photographs. I would like to thank Holly Reed of the Still Pictures Branch of the U.S. National Archives in College Park, Maryland, for a photograph of the Three Mile Island nuclear plant.

For my fifth chapter, I would like to thank Daniel Barbiero of the National Academy of Sciences and Carol Heard of the National Air and Space Museum for assistance with important documentation on the jet engine and Frank Whittle. Karyn French and Laura Gardner of the Institution of Mechanical Engineers in London were very helpful in supplying photographs of Frank Whittle and the first Whittle jet engine, along with Amy Robinson of the University for the Creative Arts in Farnham, England. Neera Puttapipat of the Imperial War Museums, Monroe Conner of the NASA Armstrong Flight Research Center, and Carleton Wilkerson of the Boeing Image Archives provided valuable photographs related to early jet aircraft. I am indebted to Fordyce Williams of the Clark University Archives for a photograph of Robert Goddard, and to David P. Stern of NASA for explanation of Goddard's early experiments. The NASA History Office and the NASA Marshall Space Flight Center provided valuable scholarship and photographs on rockets and spacecraft for this chapter.

For my sixth chapter, on the transistor, I want to give special thanks to Sheldon Hochheiser, Archivist of the AT&T History Center and Archives, and to his colleague William D. Caughlin, for their assistance and provision of valuable photographs. I am grateful to Andrew Begley of the National Archives in Philadelphia for a radar image, and to the website of Don Pies for the Regency TR-1 radio manual. I am also deeply indebted to the scholarship of Michael Riordan and Lillian Hoddeson on the transistor.

For my seventh chapter, on the microchip, in addition to the Kilby family, I am indebted to Ed Millis for his biography of Jack Kilby. I would also like to give special thanks to Professor William Case of Grinnell College for his support of my father's work and for supplying rare photographs of Robert Noyce as a student. I would like to thank Allison Haack and the Grinnell College Library Special Collections and Archives and the Grinnell College Office of Communication for permission to publish one of these photographs of Noyce. I would like to thank Texas Instruments Corporation for permission to include company photographs of Jack Kilby and his innovations, the DeGolyer Library at Southern Methodist University for information from their Kilby papers, and Michael Shulman of Magnum Photos for a photograph of the Fairchild founders. Fairchild Semiconductor Corporation supplied a photograph of the first commercial integrated circuit. I would also like to thank Daniel Hartwig, Archivist of Stanford

University, for a photograph of Frederick Terman, and Barbara Natanson of the Library of Congress for help with my Kilby research. T. R. Reid's short book, *The Chip*, was vital to my grasp of how Kilby and Noyce made their contributions to the integrated circuit or microchip, and I benefited from Leslie Berlin's biography of Robert Noyce.

For my last chapter, on the computer, I am grateful to the Seeley Mudd Archives at Princeton University and to Stephen Ferguson, Christa Cleeton, and Chloe Pfendler for a graduate student photograph of Alan Turing; and to Daniel Linke and Patricia McGuire, Archivists of Princeton and of King's College, Cambridge, respectively, for advice on the provenance of the photograph. I would also like to thank Michael John Muuss and the U.S. Army Research Laboratory for photographs of the ENIAC computer; and Dawn Stanford and Max Campbell of the IBM Corporate Archives, and International Business Machines Corporation, for photographs and permission to use them. I am grateful to the Hagley Museum and Library, and to Brian C. Daly of Unisys Corporation, for an image of Grace Murray Hopper, and I am also grateful to Douglas Misner and Greg Walz of the Utah State Historical Society Research Center for a photograph of Philo Farnsworth. I would like to thank Michael Holley for a photograph of the Altair 8800 microcomputer. Getty Images supplied photographs of Steve Wozniak, Bill Gates, and Claude Shannon. I am grateful to Susie Mulhern of the Palo Alto Research Center, a Xerox Company, for a photograph of children testing the Alto computer, and to Beth Bernstein of the RAND Corporation for a diagram by Paul Baran of distributed communication. I would also like to thank the CERN Press Office in Geneva, Switzerland, for a photograph of Timothy Berners-Lee. The Computer History Museum was invaluable in my research on the computer, and for their assistance I would like to thank Sydney Gulbronson, Sarah Lott, Massimo Petrozzi, and Carina Sweet.

Finally, this book would not have been published without the dedicated effort of Jermey Matthews of the MIT Press, assisted by Gabriela Bueno Gibbs, Haley Biermann, Deborah Cantor-Adams, Stephanie Sakson, Mary Reilly, Marge Encomienda, Susan Clark, Jim Mitchell, Molly Grote, and Heather Goss.

I want to thank Princetonians Judy McCartin Scheide for her hospitality and interest in the work that my father did with me, and Terry Towe for encouragement and advice. Eric Henney and Deborah Tegarden urged me to move forward with the book, as did many friends including Glenn Speer, Aaron Trehub, Aaron Forsberg, Jennifer Loehlin, W. Travis Hanes, James K. Chiu, Andrew Browning, and Jennie Wadsworth.

The Flanders family (Stephen and Hedy, Jefferson and Maisie, Tony and Bunny, Julie and Emil, and Carl and Andrea), the Laity family (Jim and Mary Ann, Susan, Kate and Earl, Bill, and John), and the Black family (Jim and Martha, and Christina) have shared much of my journey through life. I am grateful for their friendship and support.

I owe a special debt to the encouragement of Neva Wing, Carol Flanders, Corinne Black, and Mary Laity. I am also deeply indebted to my Billington and Bergquist grandparents, Nelson and Jane, and Jonathan and Gerda; to my step-grandmother Dorothy Gaunt Billington and to Bill and Wendy Gaunt; and to my uncles and aunts: James and Marjorie Billington, John and Lynn Billington, Janet and John Fisher, Arloa Bergquist, and to all of my Billington and Bergquist cousins.

My deepest thanks go to my immediate family—my sisters and brothers Elizabeth and Donald, Jane and Johnson, Philip and Ninik, Stephen and Miriam, and Sarah and Peter—for their love and support and for giving me eleven nieces and nephews. To my parents, David Perkins Billington, Sr., and Phyllis Bergquist Billington, I am indebted most of all, and to their memory I have dedicated this book.

Notes

Preface

1. David P. Billington, *The Innovators: The Engineering Pioneers Who Made America Modern* (New York: John Wiley and Sons, 1996).

2. David P. Billington and David P. Billington, Jr., *Power Speed and Form: Engineers and the Making of the Twentieth Century* (Princeton, NJ: Princeton University Press, 2006).

3. See Vaclav Smil, *Creating the Twentieth Century: Technical Innovations of 1867 to 1914 and Their Lasting Impact* (New York: Oxford University Press, 2005) and *Transforming the Twentieth Century: Technical Innovations and Their Consequences* (New York: Oxford University Press, 2006). For the survey commissioned by the National Academy of Engineering, see George Constable and Bob Somerville, *A Century of Innovation: Twenty Engineering Achievements That Transformed Our Lives* (Washington, DC: Joseph Henry Press, 2003).

Introduction

1. Electricity came to be used either as a source of power (e.g., for lighting) or as a new means of communication (e.g., wireline telephony). Internal combustion was the combustion of fuel to drive mechanical action, in one or more combustion chambers inside an engine. Steam engines used external combustion, in which fuel burned in a boiler to produce steam. The steam then went into a separate driving mechanism.

2. For the role of science at the inception of major innovations, see Harold C. Passer, "Electrical Science and the Early Development of the Electrical Manufacturing Industry in the United States," *Annals of Science* 7, no. 4 (December 28, 1951): 383–429; David A. Hounshell, "Two Paths to the Telephone," *Scientific American* 244 (January 1981): 156–163; Lynwood Bryant, "The Origin of the Automobile Engine," *Scientific American* 216 (March 1967): 102–113; John D. Anderson, Jr., *A History of Aerodynamics and Its Impact on Flying Machines* (Cambridge: Cambridge University Press, 1997), 192, 242–243; and Sungook Hong, "Marconi and the Maxwellians: The Origins of Wireless Telegraphy Revisited," *Technology and Culture* 35, no. 4 (October 1994): 717–749. The exception was the chemical industry, in which laboratory science played a central role. The

discoveries earlier in the nineteenth century of electrical principles by Ohm, Faraday, Henry, and others were as much principles of electrical engineering as principles of science. They are considered science because their discoverers had no practical motive in mind.

3. This is one of many useful points in W. Brian Arthur, *The Nature of Technology: What It Is and How It Evolves* (New York: Simon & Schuster/Free Press, 2009).

4. Some later writers reacted to Edison's self-promotion by denying that he was a modern engineer, calling him a mere inventor instead. For Edison, see David P. Billington and David P. Billington, Jr., *Power Speed and Form: Engineers and the Making of the Twentieth Century* (Princeton, NJ: Princeton University Press, 2006), 17–25. For the brilliance of the Wright brothers, see Howard S. Wolko, ed., *The Wright Flyer: An Engineering Perspective* (Washington, DC: Smithsonian Institution Press, 1987).

5. See Walter G. Vincenti, *What Engineers Know and How They Know It: Analytical Studies from Aeronautical History* (Baltimore, MD: Johns Hopkins University Press, 1990), 7–9.

6. For the expert view of electrical power distribution in the 1870s, see Paget Higgs, *The Electric Light in Its Practical Application* (London: E. and F. L. Spon, 1879), 158–175. For Edison's response, see Billington and Billington, *Power Speed and Form*, 220–222. Thomas Edison did not invent the incandescent light bulb (Sir Joseph Swan invented a low-resistance bulb in the 1840s). What Edison designed was a high-resistance bulb that would make a local system to distribute electric power to lamps efficient and economical, which low-resistance bulbs could not do.

Chapter 1

1. For the urban and rural populations of the United States in 1920, see the *Statistical Abstract of the United States 1930* (Washington, DC: U.S. Government Printing Office, 1930), 45–46. Populations living in communities of 2,500 inhabitants or more were classified as urban. Almost 44 percent lived in places of 8,000 or more; see ibid., 47.

2. For the industries of the mid-nineteenth century, see David P. Billington, *The Innovators: The Engineering Pioneers Who Made America Modern* (New York: John Wiley and Sons, 1996); and James C. Williams, "The American Industrial Revolution," in *A Companion to American Technology*, ed. Carroll Pursell (Oxford: Blackwell Publishing, 2005), 31–51. For the industries of the late nineteenth and early twentieth centuries, see again David P. Billington and David P. Billington, Jr., *Power Speed and Form: Engineers and the Making of the Twentieth Century* (Princeton, NJ: Princeton University Press, 2006); and Vaclav Smil, *Creating the Twentieth Century: Technical Innovations of 1867–1914 and Their Lasting Impact* (New York: Oxford University Press, 2005).

3. For the development of local water supplies, see Nelson M. Blake, *Water for the Cities: A History of the Urban Water Supply Problem in the United States* (Syracuse, NY: Syracuse University Press, 1956); and Martin Melosi, *The Sanitary City: Urban Infrastructure in America from Colonial Times to the Present* (Baltimore, MD: Johns Hopkins University Press, 2000). For the civil engineering of public water works at this time, see F. E. Turneaure and H. L. Russell, *Public Water Supplies: Requirements, Resources, and Construction of Works*, 4th ed. (New York: John Wiley and Sons, 1940).

4. See A. Wolman and L. H. Enslow, "Chlorine Absorption and Chlorination of Water," *Journal of Industrial and Engineering Chemistry* 11 (1919): 206–213. For Wolman, see M. Gordon Wolman, "Abel Wolman 1892–1989," in National Academy of Sciences, *Biographical Memoirs* 83 (2003): 3–18.

5. For an overview of America's rivers, see Arthur C. Benke and Colbert E. Cushing, eds., *Rivers of North America* (New York: Elsevier/Academic Press, 2005). The bibliographies appended to each chapter provide further information on the natural ecosystems and the impact of human activity on the rivers.

6. On the early navigation of American rivers, see Louis C. Hunter, *Steamboats on the Western Rivers: An Economic and Technological History* (Cambridge, MA: Harvard University Press, 1949). For the engineering of the steamboat and the use of water to power the early textile industry, see Billington, *The Innovators*, 41–94. On the use of Niagara Falls to generate electricity, see R. Belfield, "The Niagara System: The Evolution of an Electric Power Complex at Niagara Falls, 1883–1896," *Proceedings of the IEEE* 64, no. 9 (September 1976): 1344–1350.

7. For the U.S. Army Corps of Engineers in the nineteenth century, see Todd Shallat, *Structures in the Stream: Water, Science, and the Rise of the U.S. Army Corps of Engineers* (Austin: University of Texas Press, 1994). On an early lock and dam of the Corps on the Ohio River, see Leland R. Johnson, *The Davis Island Lock and Dam, 1870–1922* (Pittsburgh, PA: U.S. Army Engineer District, 1985).

8. For the Reclamation Service, later the Bureau of Reclamation, see Donald J. Pisani, *Water and American Government: The Reclamation Bureau, National Water Policy, and the West, 1902–1935* (Berkeley: University of California Press, 2002).

9. For a history of federal dams mainly west of the Mississippi River, see David P. Billington and Donald C. Jackson, *Big Dams of the New Deal Era: A Confluence of Engineering and Politics* (Norman: University of Oklahoma Press, 2006); and David P. Billington, Donald C. Jackson, and Martin V. Melosi, *The History of Large Federal Dams: Planning, Design, and Construction in the Era of Big Dams* (Denver, CO: U.S. Department of the Interior, Bureau of Reclamation, 2005). For the Colorado River before the era of major dams, see Paul L. Kleinsorge, *The Boulder Canyon Project: Historical and Economic Aspects* (Stanford, CA: Stanford University Press, 1941), 2–15. For the discharge of the Colorado River before Hoover Dam, see Jay Kammerer, *Water Fact Sheet: Largest Rivers in the United States* (Washington, DC: U.S. Geological Survey, 1990), 1. For the Mississippi, see ibid., 2. The Colorado River discussed in this chapter is separate from the smaller Colorado River that flows through Texas to the Gulf of Mexico.

10. On the early Imperial Valley, see Norris Hundley, Jr., *Water and the West: The Colorado River Compact and the Politics of Water in the American West*, 2nd ed. (Berkeley: University of California Press, 2009), 17–36.

11. Northern California also needed water for its cities. In 1913, Congress authorized San Francisco to dam the Tuolumne River at Hetch Hetchy, on federal land in the Sierra Nevada mountains. A pipeline from the resulting reservoir supplied water westward to San Francisco and

surrounding counties. For the early efforts of San Francisco and Los Angeles to obtain water, see Norris Hundley, Jr., *The Great Thirst: Californians and Water, A History*, 2nd ed. (Berkeley: University of California Press, 2001), 121–202. For the population growth of Los Angeles from 1900 to 1920, see the U.S. Bureau of the Census, *Fourteenth Census of the United States Taken in the Year 1920*, 12 vols. (Washington, DC: U.S. Government Printing Office, 1921), 1: 76. For the population of Los Angeles in 1930, see the *Statistical Abstract of the United States 1940* (Washington, DC: U.S. Government Printing Office, 1941), 3.

12. For the Colorado Compact, see Norris Hundley, Jr., *Water and the West*, 53 et seq. For the text of the compact, see ibid., 353–359. On the Central Arizona Project, see Rich Johnson, *The Central Arizona Project, 1918–1968* (Tucson: University of Arizona Press, 1977). In 1987, southern California agencies repaid the loan from the U.S. Treasury to build Hoover Dam.

13. For the creation of the Metropolitan Water District, see Billington and Jackson, *Big Dams of the New Deal Era*, 116–122; and *The Metropolitan Water District of Southern California: History and First Annual Report*, Charles A. Bissell, comp. (Los Angeles, CA: Metropolitan Water District, 1939). For the Boulder Canyon Project Act, see *United States Statutes at Large, 1927–1929*, vol. 45, Part 1 (Washington, DC: U.S. Government Printing Office, 1929), 1057–1066.

14. For the evolution of modern dam design, the competition between arch and gravity dams, and the triumph of mass over form in the 1920s and 1930s, see Billington and Jackson, *Big Dams of the New Deal Era*, 29–70.

15. For the evolution of arched gravity dams, see David P. Billington, Chelsea Honigmann, and Moira Treacy, "From Pathfinder to Glen Canyon: The Structural Analysis of Arched, Gravity Dams," in *The Bureau of Reclamation: History Essays from the Centennial Symposium*, 2 vols. (Washington, DC: U.S. Government Printing Office, 2002), 1: 249–271. For the design of Hoover Dam, see *Boulder Canyon Project Final Reports: General History and Description of the Project* (Boulder City: U.S. Bureau of Reclamation, 1948); and Donald C. Jackson, "Origins of Boulder/Hoover Dam: Siting, Design, and Hydroelectric Power," in *The Bureau of Reclamation: History Essays from the Centennial Symposium*, 1: 273–288. For John Lucian Savage (1879–1967), see Benjamin D. Rhodes, "From Cooksville to Chunking: The Dam-Designing Career of John L. Savage," *Wisconsin Magazine of History* 72, no. 4 (summer 1989): 242–272. For Frank E. Weymouth (1874–1941), see the finding aid to his papers at the Honnold/Mudd Library, Claremont, California.

16. Billington and Jackson, *Big Dams of the New Deal Era*, 36–41.

17. For the St. Francis Dam disaster, see Norris Hundley, Jr., and Donald C. Jackson, *Heavy Ground: William Mulholland and the St. Francis Dam Disaster* (Berkeley: University of California Press, 2016).

18. For a narrative account of Hoover Dam, see Joseph E. Stevens, *Hoover Dam: An American Adventure* (Norman: University of Oklahoma Press, 1988). For an engineering history of the dam, see U.S. Department of the Interior, Bureau of Reclamation, *Boulder Canyon Project Final Reports, Part IV—Design and Construction* (Denver: U.S. Department of the Interior, 1941). This volume has a second subtitle, "Bulletin 1: General Features." For Hoover Dam, see also Billington and Jackson, *Big Dams of the New Deal Era*, 127–144; and Kleinsorge, *The Boulder Canyon Project*, 185–230.

19. On the Six Companies, see Stevens, *Hoover Dam*, 34–45. For the inflation adjusted cost, see the U.S. Bureau of Labor Statistics consumer price index inflation calculator online at: http://www.bls.gov/data/inflation_calculator.htm (link retrieved October 26, 2019). Conversion is approximate and may reflect the plunge in nominal prices after 1929. For Francis Trenholm Crowe, see the memoir on his life in *Transactions of the American Society of Civil Engineers* 113 (1948): 1397–1403 (hereafter cited as *ASCE Transactions*). See also Stevens, *Hoover Dam*, 36–46; and Al M. Rocca, *America's Master Dam Builder: The Engineering Genius of Frank T. Crowe* (New York: University Press of America, 2001).

20. For the Frank Crowe quote, see "The Earth Movers I," *Fortune Magazine* 28, no. 8 (August 1943): 103.

21. On Walker Young, see his entry in *Who's Who in Engineering: A Dictionary of the Engineering Profession*, 6th ed. (New York: Lewis Historical Publishing, 1948), 2246 (hereafter cited by year as *Who's Who in Engineering*). On Young's relationship to Crowe at the dam site, see "The Dam," *Fortune Magazine* 8, no. 3 (September 1933): 74–88.

22. For the project completion date, see Stevens, *Hoover Dam*, 34, and on the penalty for lateness in finishing the diversion tunnels, see ibid., 59. Crowe was to receive a share in the profit of the Six Companies on the project (2.5 percent), see ibid., 252. For the harsh working conditions and the strike, see ibid., 59–79.

23. For the worker's quote, see Andrew J. Dunar and Dennis McBride, *Building Hoover Dam: An Oral History of the Great Depression* (Reno and Las Vegas: University of Nevada Press, 1993), 95 (interview with Saul "Red" Wixson).

24. Injuries from gasoline engine exhaust in the tunnels led to lawsuits against the Six Companies. See Stevens, *Hoover Dam*, 206–214. On the "high scalers," workers who descended down the canyon walls, chipping away loose stone, see ibid., 103–107. Some of the high scalers were Native Americans. Workers on the ground level included a crew of African Americans. Nonwhite workers were segregated from the others.

25. For Roosevelt's vision, including employment in public works, see his first Inaugural Address, reprinted in Franklin D. Roosevelt, *Looking Forward* (New York: John Day, 1933; reprint ed., Simon & Schuster, 2009), 219–226. For the name change to Boulder Dam, see Stevens, *Hoover Dam*, 173–174.

26. On excavation of the foundations, see Stevens, *Hoover Dam*, 185–190. For the setting of the concrete, see ibid., 191–197. The release of heat when water and cement react was known as heat of hydration.

27. For Crowe's cableways, see Stevens, *Hoover Dam*, 196–197; and for completion of the dam, ibid., 191–241. For the cubic yards of concrete in the dam, see Water and Power Resources Service, *Project Data* (Washington, DC: U.S. Department of the Interior, 1981), 84. The Bureau of Reclamation changed its name to Water and Power Resources Service in 1979 but resumed its prior name in 1981. For the dedication of the dam, see Stevens, *Hoover Dam*, 243–248.

28. For the planning of the Colorado River aqueduct, see F. E. Weymouth, "Colorado Aqueduct," *Civil Engineering* 1, no. 5 (February 1931): 371–376. For the projects on the Colorado River downstream, see Kleinsorge, *The Boulder Canyon Project*, 230–244; and David Carle, *Introduction to Water in California* (Berkeley: University of California Press, 2004), 110–115. For the completion of the powerhouses and ancillary work at Hoover Dam, see Stevens, *Hoover Dam*, 248–252. Turbines or bladed shafts rotated by moving water could generate electricity (see chapter 2, boxes 2.1, 2.2, and 2.3). For the name change back to Hoover Dam, see *United States Statutes at Large*, vol. 61, Part 1 (Washington, DC: Government Printing Office, 1948), 56–57.

29. For the Central Valley Project, see Billington and Jackson, *Big Dams of the New Deal Era*, 253–275; Hundley, *The Great Thirst*, 203–275; and Carle, *Introduction to Water in California*, 103–109.

30. For the design and construction of Shasta and Friant Dams, see Billington and Jackson, *Big Dams of the New Deal Era*, 275–288. On the building of Shasta Dam, see also Pacific Constructors, *Shasta Dam and Its Builders* (San Francisco, CA: Schwabacher-Frey, 1945).

31. For the results of the Central Valley Project, see Billington and Jackson, *Big Dams of the New Deal Era*, 288–292. At the end of the twentieth century, California produced half of the nation's vegetables and more than half of its fruit. See the *1997 Census of Agriculture*, 3 vols. (Washington, DC: U.S. Department of Agriculture, 1997), vol. 2, Part 2, 37–38. For the changes in California's farm population, see Hundley, *The Great Thirst*, 240–242, 260–272. The limitation to 160 acres was for individual farmers; a married couple could have 320 acres. A common way for large farms to comply with the limitation on acreage and evade its intent was to divide large holdings among additional relatives, employees, or shareholders. In 1982 Congress raised the limit on individual holdings to 900 acres. For changes in land tenure in California after World War II, see Ellen Liebman, *California Farmland: A History of Large Agricultural Landholdings* (Totowa, NJ: Rowman and Allanheld, 1983), 129–173. For a retrospective on the project, see Erik A. Stene, "The Central Valley Project: Controversies Surrounding Reclamation's Largest Project," in *The Bureau of Reclamation: History Essays from the Centennial Symposium*, 2: 503–521.

32. For the California State Water Project, see Hundley, *The Great Thirst*, 276–302; and Carle, *Introduction to Water in California*, 92–103. For the population growth from 1920 to 1970, see the *Statistical Abstract of the United States for 1980* (Washington, DC: U.S. Bureau of the Census, 1980), 10. For an overview of the federal role in the post-1945 economic development of California and the West, see Gerald D. Nash, *The Federal Landscape: An Economic History of the Twentieth-Century West* (Tucson: University of Arizona Press, 1999).

33. For the discharge at the mouth of the Columbia today, see again Kammerer, *Water Fact Sheet*, 2. The discharge of the Colorado River was 22,000 cubic feet per second before 1934; see ibid., 1. On engineering interest in the Columbia River, see Billington and Jackson, *Big Dams of the New Deal Era*, 152–156.

34. On building the two dams and the controversy over the cost and height of Grand Coulee Dam, see Billington and Jackson, *Big Dams of the New Deal Era*, 168–170, 174–178.

35. On Bonneville Dam, see Billington and Jackson, *Big Dams of the New Deal Era*, pp. 156–170; and William F. Willingham, *Water Power in the "Wilderness": The History of Bonneville Lock and*

Dam (Portland, OR: U.S. Army Corps of Engineers, Portland District, 1997). On George Gerdes (1900–1972), see Thornton Corwin, "Henry George Gerdes," *ASCE Transactions* 139 (1974): 565. For the design features at Bonneville, see Abbie Liel and David P. Billington, "Engineering Innovation at Bonneville Dam," *Technology and Culture* 49, no. 3 (July 2008): 727–751. Hardening the downstream base and adding baffles caused overflowing water to jump up, lessening the impact. On a smaller scale, Arthur Morgan designed the outlets of Miami Conservancy dams with similar features (see chapter 2). Silica added to concrete was known to improve its water resistance and cool the concrete as it set; at Bonneville, instead of piping cold water, engineers added silica to the concrete. Engineers at Bonneville Dam installed new turbines with adjustable blades (Kaplan turbines) that managed the variable flow of water better than turbines with fixed blades used at other dams, where the flow varied less.

36. On Grand Coulee Dam, see Billington and Jackson, *Big Dams of the New Deal Era*, 165–188; and Paul C. Pitzer, *Grand Coulee: Harnessing a Dream* (Pullman: Washington State University Press, 1994). On Frank Arthur Banks (1883–1957), see his entry in *The National Cyclopaedia of American Biography*, F, 1939–1942 (New York: James T. White and Company, 1942), 450–451; and *Civil Engineering* 28, no. 2 (February 1958): 131. Like Frank Crowe, Banks came from Maine and attended its state university.

37. For the wartime contribution of the Columbia dams, see Vernon M. Murray, "Grand Coulee and Bonneville Power in the National War Effort," *Journal of Land and Public Utility Economics* 18, no. 2 (May 1942): 134–139. Alternating current reversed direction many times per second, while direct current flowed in one direction only.

38. On the Columbia River Control Plan, see Billington and Jackson, *Big Dams of the New Deal Era*, 192–199; and William Whipple, Jr., "Comprehensive Plan for the Columbia Basin," *ASCE Transactions* 115 (1950): 1426–1436. On William Whipple, Jr. (1909–2007), see William Whipple, Jr., *Autobiography* (Princeton, NJ: privately printed, 1996). Copy in the Engineering Library, Princeton University.

39. On the BPA, see Gene Tollefson, *BPA: The Struggle for Power at Cost* (Portland, OR: Bonneville Power Administration, 1987); and *Power of the River: The Continuing Legacy of the Bonneville Power Administration in the Pacific Northwest* (Portland, OR: Bonneville Power Administration, 2012). For population growth in the states of the Pacific Northwest to 1970, see the *Statistical Abstract of the United States for 1980*, 10. For the output of federal power as a share of total U.S. electricity output in 1970, see ibid., 614. The federal government later created regional power marketing administrations across the United States to sell electricity produced by federal dams. These agencies are under the U.S. Department of Energy today. In addition to the BPA, they include the Western Area Power Administration, the Southwestern Power Administration, and the Southeastern Power Administration. The north central and northeastern states have fewer federal dams and do not have federal power marketing administrations.

40. On the Bonneville fishways, see William F. Willingham, *Water Power in the Wilderness*, 47–53. See also Lisa Mighetto and Wesley J. Ebel, *Saving the Salmon: A History of the U.S. Army Corps of Engineers' Efforts to Protect Anadromous Fish on the Columbia and Snake Rivers* (Seattle, WA: Historical

Research Associates, 1994). For a critical view of the impact of dams on the Columbia River, see Blaine Harden, *A River Lost: The Life and Death of the Columbia* (New York: W. W. Norton, 1996). On the decline in fish populations, see also Joseph Taylor, *Making Salmon: An Environmental History of the Northwest Fisheries Crisis* (Seattle: University of Washington Press, 1999). Taylor notes that changing ocean conditions and failed efforts to breed the fish also contributed to the decline.

41. On the origins and planning of Fort Peck Dam, see Billington and Jackson, *Big Dams of the New Deal Era*, 200–207. Lieutenant Colonel Richard C. Moore oversaw the design and Major Thomas B. Larkin supervised construction. For the hydraulic-fill method of construction, see ibid., 207–215; and for the spillway, ibid., 215–219. The spillway gates became symbols of the New Deal when a photograph of them, taken by Margaret Bourke-White, appeared on the cover of the first issue of *Life* magazine on November 23, 1936.

42. For the quote from surveyor Ray Kendall, see Billington and Jackson, *Big Dams of the New Deal Era*, 222.

43. On the 1938 slide and its aftermath, see Billington and Jackson, *Big Dams of the New Deal Era*, 220–230. See also T. A. Middlebrooks, "Fort Peck Slide," *ASCE Transactions* 107 (1942): 723–764.

44. On the "Pick-Sloan Plan" and further dam building along the Missouri, see Billington and Jackson, *Big Dams of the New Deal Era*, 230–252. See also John R. Ferrell, *Big Dam Era: A Legislative and Institutional History of the Pick-Sloan Missouri Basin Program* (Omaha, NE: U.S. Army Corps of Engineers, 1993).

45. For tensions over Missouri River water, see John E. Thorson, *River of Promise, River of Peril: The Politics of Managing the Missouri River* (Lawrence: University Press of Kansas, 1994). See also A. Dan Turlock, "The Missouri River: The Paradox of Conflict without Scarcity," *Great Plains Natural Resources Journal* 2 (1997): 1–12. For population growth, see the *Statistical Abstract of the United States for 1980*, 10. The region is taken here to include Montana, North and South Dakota, Minnesota, Iowa, Nebraska, Kansas, and Missouri. Northeastern Wyoming would add a small number. For the changes in agriculture across the Midwest, see *The American Midwest: An Interpretive Encyclopedia*, ed. Andrew R. L. Cayton, Richard Sisson, and Christian Zacher (Bloomington: Indiana University Press, 2007), 142–145. Nationally, the productivity of farming in food crops increased more than fourfold between 1950 and 1970. In vegetable production it increased only twofold and in fruit by only one-half. See the *Statistical Abstract of the United States for 1980*, 709. The greater gains in food crops were the result of easier mechanization.

46. The Corps of Engineers submitted cost estimates for undertaking the surveys, which appeared as "Estimate of Cost of Examinations, etc. of Streams Where Power Development Appears Feasible," U.S. House of Representatives, 69th Congress, 1st Session, Document No. 308, December 7, 1925, to November 10, 1926. The subsequent surveys of the Columbia and Missouri Rivers appeared as "Columbia River and Minor Tributaries," U.S. House of Representatives, 72nd Congress, 2nd Session, Document No. 103, March 29, 1932; and "Missouri River," U.S. House of Representatives, 73rd Congress, 2nd Session, Document No. 238, February 5, 1934. On the 308 Reports, see Billington and Jackson, *Big Dams of the New Deal Era*, 85–88.

47. For the public debate over dams and electric power, see Billington and Jackson, *Big Dams of the New Deal Era*, 122–125 for Hoover Dam, and ibid., 189–192 for the Columbia River. See also Stevens, *Hoover Dam*, 26–27; and Sarah S. Elkind, "Private Power at Boulder Dam: Utilities, Government Power, and Political Realism," in *The Bureau of Reclamation: History Essays from the Centennial Symposium*, 2: 447–465. Controversy over the federal role in the Columbia dams centered on whether there should be a federal river basin authority to manage them. The Bonneville Power Administration had a more limited role as a result of local opposition to an agency with broader powers, and the Bureau of Reclamation and the Corps of Engineers opposed a stronger rival agency. See Billington and Jackson, *Big Dams of the New Deal Era*, 189–192.

48. For the Native American lands submerged by Grand Coulee Dam, see Pitzer, *Grand Coulee*, 219–222. For those at Garrison Dam, see Billington and Jackson, *Big Dams of the New Deal Era*, 239–241; and Michael L. Lawson, *Dammed Indians: The Pick-Sloan Plan and the Missouri River Sioux, 1944–1980* (Norman: University of Oklahoma Press, 1994). For those at Shasta Dam, see Bradley L. Garrett, "Drowned Memories: The Submerged Places of the Winnemem Wintu," *Archaeologies: Journal of the World Archaeological Congress* 6, no. 2 (2010): 346–371.

49. For U.S. electric generating capacity in the 1930s and the share supplied by hydroelectric power, see U.S. Department of Commerce/Bureau of the Census, *Statistical Abstract of the United States 1940* (Washington, DC: U.S. Government Printing Office, 1941), 401. For the initial generating capacity at Hoover Dam, see Kleinsorge, *The Boulder Canyon Project*, 281–282. In 1998, hydroelectric power supplied 9.6 percent of the net generation of electricity in the United States. See U.S. Bureau of the Census, *Statistical Abstract of the United States 2000* (Washington, DC: U.S. Government Printing Office, 2001), 593.

50. For the failure of the Oroville Dam spillway, see John W. France et al., *Independent Forensic Team Report: Oroville Dam Spillway Incident* (Sacramento, CA: California Department of Water Resources, January 5, 2018).

51. For the problem of siltation in Lake Mead, see W. O. Smith, C. P. Vetter, G. B. Cummings, et al., *Comprehensive Survey of Sedimentation in Lake Mead, 1948–49*, U.S. Geological Survey Professional Paper No. 295 (Washington, DC: U.S. Government Printing Office, 1960). The authors argued that siltation would take four centuries to reach the crest of Hoover Dam but noted that the rate of buildup was faster than expected. The estimate came before the building of Glen Canyon Dam upstream.

52. For Glen Canyon Dam, see Water and Power Resources Service, *Project Data* (Washington, DC: U.S. Department of the Interior, 1981), 355, 361–363. The Glen Canyon Dam is 710 feet (216 m) high with a base measuring 300 feet (91 m) from front to back and a crest length of 1,560 feet (475 m). On the controversy over the Green River dams, see Mark Harvey, *A Symbol of Wilderness: Echo Park and the American Conservation Movement* (Albuquerque: University of New Mexico Press, 1994). On Glen Canyon, see Eliot Porter, *The Place No One Knew: Glen Canyon on the Colorado* (San Francisco, CA: Sierra Club, 1963); Jared Farmer, *Glen Canyon Dammed: Inventing Lake Powell and the Canyon Country* (Tucson: University of Arizona Press, 1999); and W. L. Rusho, "Bumpy Road for Glen Canyon Dam," in *The Bureau of Reclamation: History Essays from the Centennial Symposium*, 2: 523–549. The classic indictment of modern dam-building in the West is Mark Reisner, *Cadillac Desert: The American West and Its Disappearing Water* (New York: Viking Books, 1986).

Chapter 2

1. For Morgan's life, see Aaron D. Purcell, *Arthur Morgan: A Progressive Vision for American Reform* (Knoxville: University of Tennessee Press, 2014), 11–37. See also Roy Talbert, *FDR's Utopian: Arthur Morgan of the TVA* (Jackson: University Press of Mississippi, 1994), 1–21. Talbert provides a more critical view of Morgan and his career. Purcell gives more detail about Morgan's life and thought after the TVA.

2. For Morgan's apprenticeship and success as a drainage expert, see Talbert, *FDR's Utopian*, 22–35.

3. On the Dayton flood, the organization of the District, and Morgan's role, see Arthur E. Morgan, *The Miami Conservancy District* (New York: McGraw-Hill, 1951), 11–203. Morgan prints an excerpt from *Frank Leslie's Illustrated Weekly Newspaper*, April 24, 1913, describing how Orville Wright narrowly avoided the loss of his workshop and papers during the flood. See Morgan, *The Miami Conservancy District*, 36. On The Miami Conservancy, see also Purcell, *Arthur Morgan*, 39–40, 65–69, 82–93.

4. For Ellet's report, see Charles Ellet, *Report on the Overflows of the Delta of the Mississippi River*, Senate Executive Document No. 20, 32nd Congress, 1st Session (Washington, DC: U.S. Senate, 1852).

5. For the Humphreys report, see A. A. Humphreys and H. L. Abbot, *Report upon the Physics and Hydraulics of the Mississippi River . . .* (Philadelphia: J. B. Lippincott, 1861). On the controversy over how to control flooding, and the failure of the "levees only" policy, see Martin Reuss, "Andrew A. Humphreys and the Development of Hydraulic Engineering: Politics and Technology in the Army Corps of Engineers, 1850–1950," *Technology and Culture* 26, no. 1 (January 1985): 1–33.

6. For Morgan's plan, see Arthur E. Morgan, *Report of the Chief Engineer*, 3 vols. (Dayton, OH: The Miami Conservancy District, 1916); and Morgan, *The Miami Conservancy District*, 204–274. Morgan benefited from the ability of Gordon Rentschler, later chairman of the First National City Bank of New York (now Citibank) and a leading citizen in nearby Hamilton, Ohio, to raise the $35 million to pay for the conservancy project.

7. Quote is from Morgan, *The Miami Conservancy District*, 284.

8. For Morgan's approach and the idea to build basins with open conduits, see Morgan, *The Miami Conservancy District*, 284–286. On the construction program and its completion, see ibid., 312–385, 405–426. See also William G. Hoyt and Walter B. Langbein, *Floods* (Princeton, NJ: Princeton University Press, 1955), 228–230. Morgan also built some levees along the Miami River itself.

9. On Morgan's engineering organization and his relations with labor, see Morgan, *The Miami Conservancy District*, 275–311, 386–404. For the farms and parkland, see ibid., 346–349.

10. On the Moraine Park School, which enjoyed the support of Orville Wright and other local figures, see Purcell, *Arthur Morgan*, 79–82. In 1920, Morgan became the first president of the

Association for the Advancement of Progressive Education, a group whose vice presidents were the educator John Dewey and the British writer H. G. Wells. For Morgan's involvement in the Association, which was limited, and for his work at Antioch College, see Talbert, *FDR's Utopian*, 44–68; and Purcell, *Arthur Morgan*, 99–120, 124–133. The primary school, The Antioch School, is now independent but continues to involve pupils in its governance. The Arthur Morgan School, near Asheville, North Carolina, is a middle school today organized along similar lines.

11. On contact with Eleanor Roosevelt, see Morgan, *The Making of the TVA* (Buffalo, NY: Prometheus Books, 1974), 8, 175. However, Talbert is less sure that Mrs. Roosevelt was an influence on Morgan's later appointment. See Talbert, *FDR's Utopian*, 82. For Franklin Roosevelt's interest in Antioch College, see ibid., 84–86. For Morgan's appointment, see Purcell, *Arthur Morgan*, 135–142.

12. On conditions in the Tennessee Valley, see Joseph Sirera Ransmeier, *The Tennessee Valley Authority: A Case Study in the Economics of Multiple Purpose Stream Planning* (Nashville, TN: Vanderbilt University Press, 1942), 82–90.

13. On Wilson Dam and the debate over its disposition, see Preston J. Hubbard, *Origins of the TVA: The Muscle Shoals Controversy* (Nashville, TN: Vanderbilt University Press, 1961), 1–27. See also Margaret Jackson Clarke, "The Federal Government and the Fixed Nitrogen Industry, 1915–1926," PhD dissertation (Oregon State University, 1977). For the opposition of Norris to private control, and the veto of public power by Coolidge and Hoover, see Hubbard, *Origins of the TVA*, 217–266.

14. For his critical view of the electric power industry, see Franklin Roosevelt, *Looking Forward* (New York: John Day, 1933; reprint ed., Simon & Schuster, 2009), 111–124.

15. For the isolation of rural areas from electric power, see D. Clayton Brown, *Electricity for Rural America: The Fight for the REA* (Westport CT: Greenwood Press, 1980), 3–12. For the number of Americans living on farms in 1930, see *Historical Statistics of the United States: Colonial Times to 1970* (Washington, DC: U.S. Bureau of the Census, 1975), Part 1, 12–13. For the number with electric service, see ibid., Part 2, 827.

16. The Tennessee Valley Authority Act with subsequent amendments may be found in the *United States Code*, Title 16, Section 831. For the original act, see *United States Statutes at Large, 1933–1934*, vol. 48 (Washington, DC: U.S. Government Printing Office, 1934), Part 1, 58–72. For the inspiration of the TVA, see Paul K. Conkin, "Intellectual and Political Roots," in *TVA: Fifty Years of Grass-Roots Bureaucracy*, ed. Erwin C. Hargrove and Paul K. Conkin (Urbana: University of Illinois Press, 1983), 3–34. For its early years, see Richard Lowitt, "The TVA, 1933–45," in ibid., 35–65.

17. For John Harcourt Alexander Morgan, see his entry in *Who Was Who in America, 1951–1960*, vol. 3 (Chicago, IL: A. N. Marquis, 1960). For David Eli Lilienthal, see Steven M. Neuse, *David Lilienthal: The Journey of an American Liberal* (Knoxville: University of Tennessee Press, 1996). Arthur Morgan mentions consulting Justice Louis Brandeis, who had followed Morgan's work at Antioch College with interest, for advice on who to appoint. Brandeis recommended Lilienthal.

See Morgan, *The Making of the TVA*, 22. For Lilienthal's work in the TVA, see also Erwin C. Hargrove, "David Lilienthal and the Tennessee Valley Authority," in *Leadership and Innovation: A Biographical Perspective on Entrepreneurs in Government*, ed. Jameson W. Doig and Erwin C. Hargrove (Baltimore, MD: Johns Hopkins University Press, 1987), 25–60.

18. For the organization of the TVA, see Morgan, *The Making of the TVA*, 18–37; and on its dams, see ibid., 93–103. See also C. Herman Pritchett, *The Tennessee Valley Authority: A Study in Public Administration* (Chapel Hill: University of North Carolina Press, 1943), 147–183.

19. For the Corps of Engineers surveys and the initial work on Cove Creek, see David P. Billington and Donald C. Jackson, *Big Dams of the New Deal Era: A Confluence of Engineering and Politics* (Norman: University of Oklahoma Press, 2006), 85–88, 91–92. For the work of the Corps, see also Leland R. Johnson, *Engineers on the Twin Rivers: A History of the Nashville District* (Nashville, TN: U.S. Army Engineer District, 1978), 181–184. For his critique of the Corps over its "levees-only" policy at the time, see Arthur E. Morgan, "The Basis of the Case Against Reservoirs for Mississippi Flood Control," *ASCE Transactions* 93 (1929): 737–754. The other four dams were Pickwick Landing in Tennessee, Wheeler and Guntersville in Alabama, and Hiwassee in North Carolina.

20. To create a public engineering organization, rather than contract the work to private firms, was known as the "force account" method. In addition to accelerating construction time, the approach allowed Morgan to set better terms of employment for the workers. A private contractor might have held down wages and working conditions if necessary in order to clear a profit. For Morgan's organization, see A. B. Liel, "The Influence of Engineering Organization on Design and Construction Processes at Tennessee Valley Authority Dams in the 1930s," in *ASCE Structures Congress*, Las Vegas, Nevada, April 2011, 12 pp. For the personnel policy of the TVA, see also Morgan, *The Making of the TVA*, 83–85, 118–130. For the numbers of workers hired, see the *Annual Report of the Tennessee Valley Authority for the Fiscal Year Ended June 30, 1935* (Washington, DC: U.S. Government Printing Office, 1936), 46. Hereafter cited as *Annual Report of the TVA . . . for 1935*. Later annual reports are cited by year.

21. On Norris Dam, see Tennessee Valley Authority, *The Norris Project*, Technical Report No. 1 (Washington, DC: U.S. Government Printing Office, 1940). For the dimensions, see ibid., 71. For the other dams, see *The Wheeler Project*, Technical Report No. 2 (Knoxville, TN: Tennessee Valley Authority, 1940); *The Pickwick Landing Project*, Technical Report No. 3 (Knoxville, TN: Tennessee Valley Authority, 1941); *The Guntersville Project*, Technical Report No. 4 (Knoxville, TN: Tennessee Valley Authority, 1941); and *The Hiwassee Project* (Washington, DC: U.S. Government Printing Office, 1946). Summary data on all dams appeared in the *Annual Reports* of the TVA. The total expenditure of $87 million was close to 1 percent of the federal budget in 1940.

22. On the flood control and navigation improvements, see *Annual Report of the TVA for . . . 1935*, 3–5; and *Annual Report of the TVA for . . . 1940*, 3–4. For the control of malaria, see ibid., 35–36; and for land acquisition, ibid., 38–39. For a study of the residents displaced by Norris Dam, see Michael J. McDonald and John Muldowny, *TVA and the Dispossessed: The Resettlement of Population in the Norris Dam Area* (Knoxville: University of Tennessee Press, 1982).

23. For the engineering of electricity generation and distribution, see David P. Billington and David P. Billington, Jr., *Power Speed and Form: Engineers and the Making of the Twentieth Century* (Princeton, NJ: Princeton University Press, 2006), 13–31.

24. On the beginning of the power program and its outreach to rural customers, see *Annual Report of the TVA for . . . 1935*, 24–34; and for Wilson Dam and its power, see ibid., 24, 29. For power production and consumption in 1940, see *Annual Report of the TVA for . . . 1940*, 5, 19–30. A pie chart between ibid., 18–19, gives the recipients of power by category.

25. On the fertilizer program, see *Annual Report of the Tennessee Valley Authority for . . . 1935*, 18–23. For the wider impact on forestry and farming, see Tennessee Valley Authority, *Unified Valley Development: TVA Reports 1946* (Washington, DC: U.S. Government Printing Office, 1946), 5–21.

26. The Port of New York Authority is now the Port Authority of New York and New Jersey. See Jameson W. Doig, *Empire on the Hudson: Entrepreneurial Vision and Political Power at the Port of New York Authority* (New York: Columbia University Press, 2001). Sources on the Bonneville Power Administration and the Metropolitican Water District of Southern California may be found in notes 13 and 39 to chapter 1.

27. For Morgan's vision of society, see Morgan, *The Making of the TVA*, 58–59, 67–69, 87–91, 183–199; and Arthur E. Morgan, *The Small Community, Foundation for Democratic Life* (New York: Harper and Brothers, 1942). For the town of Norris, see Tennessee Valley Authority, *The Norris Project*, 173–220; and Walter L. Creese, *TVA's Public Planning: The Vision, the Reality* (Knoxville: University of Tennessee Press, 1990), 238–263.

28. For Lilienthal's vision of society and the role of public power, see David E. Lilienthal, "Business and Government in the Tennessee Valley," *Annals of the American Academy of Political and Social Science* 172 (March 1934): 45–49; and David E. Lilienthal, *TVA: Democracy on the March* (New York: Penguin Books, 1944). For standard residential electric rates and those charged by the TVA to customers in the Wilson Dam service area in 1934–1935, see *Annual Report of the Tennessee Valley Authority for . . . 1935*, 29–31. Electricity usage in the Tennessee Valley increased by twice the rate of national usage as a result of the lower TVA rates. For a larger assessment of the TVA's early electrical program, see Thomas K. McCraw, "Triumph and Irony—The TVA," *Proceedings of the IEEE* 64, no. 9 (September 1976): 1372–1380.

29. For the conflict of the TVA with private utilities, see Thomas K. McCraw, *TVA and the Power Fight, 1933–1939* (Philadelphia: Lippincott, 1971), 111–115. For the 1936 ruling, see "Ashwander et al. vs. Tennessee Valley Authority et al.," *United States Reports*, vol. 297 (Washington, DC: U.S. Government Printing Office, 1936), 288–372.

30. On the tension between Morgan and Lilienthal, see McCraw, *TVA and the Power Fight*, 91–96, 115–121, and 131–133; Thomas K. McCraw, *Morgan vs. Lilienthal: The Feud within the TVA* (Chicago, IL: Loyola University Press, 1970); and Talbert, *FDR's Utopian*, 150–168. For the 1939 Supreme Court decision, see "Tennessee Electric Power Company et al. v. Tennessee Valley Authority et al.," *United States Reports*, vol. 306 (Washington, DC: U.S. Government Printing Office, 1939), 118–152. For Wendell Willkie, see Muriel Rukeyser, *One Life* (New York: Simon &

Schuster, 1957). Willkie's running mate for vice president, Senator Charles McNary of Oregon, supported the building of Bonneville Dam.

31. For Arthur Morgan's departure from the TVA, see Talbert, *FDR's Utopian*, 169–194; and Purcell, *Arthur Morgan*, 171–203. Morgan devoted the rest of his life to Quaker religious activism and to the creation of small communities in rural areas of the United States and other countries. See ibid., 213–274. For Lilienthal's TVA legacy, see again Hargrove, "David Lilienthal and the Tennessee Valley Authority," in *Leadership and Innovation*, 25–60; and Neuse, *David E. Lilienthal*, 138–144.

32. For the numbers of African Americans employed by the TVA, see the *Annual Report of the TVA for . . . 1935*, 48. See also Cranston Clayton, "The TVA and the Race Problem," *Opportunity: A Journal of Negro Life* 12, no. 4 (April 1934): 111–112. For the broader social failings of the agency, see Richard Lowitt, "The TVA, 1933–45," in *TVA: Fifty Years of Grass-Roots Bureaucracy*, 58–60.

33. On the REA, see John M. Carmody, "Rural Electrification in the United States," *Annals of the American Academy of Political and Social Science* 201, no. 1 (January 1939): 82–88; and Brown, *Electricity for Rural America*. For the number of Tennessee Valley residents served by electricity by 1946, see the *Annual Report of the TVA for . . . 1946*, 67.

34. For Lilienthal's tenure at the Atomic Energy Commission, see Neuse, *David E. Lilienthal*, 167–228. After leaving the AEC, Lilienthal worked as a consultant and promoted economic development in other countries with the TVA as a model; see ibid., 245–286, 292–312. Dams inspired by the TVA in Africa and Asia brought electric power to these countries but did less to modernize their societies. For the TVA influence abroad, see Daniel Klingensmith, *"One Valley and a Thousand": Dams, Nationalism, and Development* (New Delhi: Oxford University Press, 2007).

35. For his argument in favor of regional authorities, see Lilienthal, *Democracy on the March*, 122–144.

36. On opposition to the TVA model elsewhere in the United States, see William E. Leuchtenberg, "Roosevelt, Norris and the Seven Little TVAs," *Journal of Politics* 14, no. 3 (August 1952): 418–441; and Craufurd D. Goodwin, "The Valley Authority Idea—The Fading of a National Vision," in *TVA: Fifty Years of Grass-Roots Bureaucracy*, 263–296.

37. See Wilmon H. Droze, "The TVA, 1945–80: The Power Company," in *TVA: Fifty Years of Grass-Roots Bureaucracy*, 66–85. See also Erwin C. Hargrove, *Prisoners of Myth: The Leadership of the Tennessee Valley Authority, 1933–1990* (Princeton NJ: Princeton University Press, 1994).

38. On the Paradise power plant, see Reed A. Elliot, Walter F. Emmons, and Henry T. Lofft, "TVA's Paradise Steam Plant," *Journal of the Power Division: Proceedings of the American Society of Civil Engineers* 88, no. 1 (May 1962): 89–119. The coal consumed at Paradise in 1970 was 2 percent of the coal and 1 percent of the energy (expressed in coal-equivalent tons) used to generate electricity in the United States that year. See *Historical Statistics of the United States: Colonial Times to 1970*, Part 2, 826. In 1972, the TVA built a second plant as large as Paradise near the Cumberland River in Kentucky.

39. On environmental controversy over the TVA in the postwar era, see William Bruce Wheeler and Michael J. McDonald, *TVA and the Tellico Dam 1936–1979: A Bureaucratic Crisis in Post-Industrial America* (Knoxville: University of Tennessee Press, 1986).

40. For the rise in costs and rates, see again Droze, "The TVA, 1945–80: The Power Company," in *TVA: Fifty Years of Grass-Roots Bureaucracy*, 78–81. The TVA had to finance its power operations by itself after 1959.

41. See Michael Wines, "T.V.A. Speeds Push Away from the Burning of Coal," *New York Times*, November 14, 2013, A18. On the new plant, see "Paradise Combined Cycle Plant," online at: https://www.tva.gov/Energy/Our-Power-System/Natural-Gas/Paradise-Combined-Cycle-Plant. In 2017, the TVA produced 37 percent of its energy from nuclear plants, 24 percent from coal, 20 percent from natural gas, 9 percent from dams generating hydroelectricity, and the rest from solar, wind, and other sources. Data for 2017 obtained online at: https://www.tva.com/About-TVA/TVA-at-a-Glance (links retrieved October 17, 2019).

42. For the flood of 1927, see John M. Barry, *Rising Tide: The Great Mississippi Flood of 1927 and How It Changed America* (New York: Simon & Schuster, 1997). For the response by the Army Corps of Engineers, see Major General Edgar Jadwin, "The Plan for Flood Control of the Mississippi River in Its Alluvial Valley," *Annals of the American Academy of Political and Social Science* 135, no. 1 (1928): 35–56. Morgan's 1929 article in the *ASCE Transactions* (cited in note 19 above) attacked this response as inadequate. On the new channels, see Martin Reuss, *Designing the Bayous: The Control of Water in the Atchafalaya Basin 1800–1995* (College Station: Texas A&M University Press, 2004); and Emory Kemp, *Stemming the Tide: Design and Operation of the Bonnet Carré Spillway* (Chicago, IL: Public Works Historical Society, 1990).

43. On the 1993 flooding in the upper Midwest, see Gerald Galloway et al., *Sharing the Challenge: Floodplain Management into the 21st Century* (Washington, DC: U.S. Government Printing Office, 1994). On Hurricane Katrina, see *Hurricane Katrina: A Nation Still Unprepared*, Special Report of the Committee on Homeland Security and Governmental Affairs, United States Senate, Together with Additional Views, 109th Congress, 2nd Session, Senate Report 109–322 (Washington, DC: U.S. Government Printing Office, 2006).

Chapter 3

1. For the 1939 World's Fair, see *Dawn of a New Day: The New York World's Fair, 1939/40*, ed. Helen A. Harrison (New York: Queens Museum/New York University Press, 1980). For the Futurama ride, see Norman Bel Geddes, *Magic Motorways* (New York: Random House, 1940). Bel Geddes, a noted architect, conceived the ride.

2. On the mass-produced car and its impact, see Robert Casey, *The Model T: A Centennial History* (Baltimore, MD: Johns Hopkins University Press, 2008); and James P. Womack, Daniel T. Jones, and Daniel Roos, *The Machine That Changed the World* (New York: Rawson Associates/Simon & Schuster, 1990), 21–38. On the oil industry, see John Lawrence Enos, *Petroleum Progress and Profits: A History of Process Innovation* (Cambridge, MA: MIT Press, 1962). For early motor vehicle

registrations, see *Automobiles: Facts and Figures 1940* (Detroit, MI: Automobile Manufacturers Association, 1940), 11.

3. For an overview of early road building in America, see Thomas H. MacDonald, "The History and Development of Road Building in the United States," *ASCE Transactions* 92 (1928): 1181–1206; and *History of Public Works in the United States 1776–1976*, ed. Ellis L. Armstrong (Chicago, IL: American Public Works Association, 1976), 53–82.

4. On the Good Roads movement, see Bruce E. Seely, *Building the American Highway System: Engineers as Policy Makers* (Philadelphia: Temple University Press, 1987), 11–23. For the engineering of the Model T, see David P. Billington and David P. Billington, Jr., *Power Speed and Form: Engineers and the Making of the Twentieth Century* (Princeton, NJ: Princeton University Press, 2006), 87–92.

5. For the beginning of aid by state governments to local road building, see MacDonald, "The History and Development of Road Building in the United States," *ASCE Transactions* 92 (1928): 1196–1197. For the Federal Aid Road Act of 1916, see *United States Statutes at Large*, vol. 39 (1915–17), Part 1, 355–359; and Seely, *Building the American Highway System*, 36–45.

6. Small quantities of additional minerals in the original cement are neglected here. For the history of Portland cement, see Robert W. Lesley, *History of the Portland Cement Industry in the United States* (Chicago, IL: International Trade Press, 1924; reprint ed., New York: Arno Press, 1972). For the engineering of concrete in the early twentieth century, see Leonard Church Urquhart and Charles Edward O'Rourke, *Design of Concrete Structures* (New York: McGraw-Hill, 1923).

7. For the use of asphalt, see I. B. Holley, Jr., "Blacktop: How Asphalt Paving Came to the Urban United States," *Technology and Culture* 44, no. 4 (October 2003): 703–733. For the equipment used in modern road building, see I. B. Holley, Jr., *The Highway Revolution 1895–1925: How the United States Got Out of the Mud* (Durham, NC: Carolina Academic Press, 2008), 119–132.

8. On the wartime transportation crisis, see Federal Highway Administration, *America's Highways 1776–1976: A History of the Federal-Aid Program* (Washington, DC: U.S. Federal Highway Administration, 1977), 97–98. For a first-hand account, see Clinton Cowen, "How Ohio Handled Important Roads Broken Down by Huge Traffic," *Public Roads* 1, nos. 6–8 (December 1918): 25–27, 28.

9. The first federal agency to oversee roads was the Office of Road Inquiry, founded in 1893, which became the Office of Public Roads in 1905 and the Bureau of Public Roads in 1915, all under the U.S. Department of Agriculture. On Thomas MacDonald, see his entry in *The National Cyclopaedia of American Biography*, G, 1943–1946 (New York: James T. White & Company, 1946), 316–317. See also Earl Swift, *The Big Roads: The Untold Story of the Engineers, Visionaries, and Trailblazers Who Created the American Superhighways* (Boston: Houghton Mifflin Harcourt, 2011), 51–61. The American Association of State Highway Officials (AASHO) was organized in 1914.

10. For MacDonald's views, see Thomas H. MacDonald, "Federal-Aid Accomplishments," *Public Roads* 3, no. 32 (December 1920): 11–15. For the more scientific emphasis in the Bureau of Public Roads in the 1920s and 1930s, see Bruce E. Seely, "The Scientific Mystique in Engineering: Highway Research at the Bureau of Public Roads, 1918–1940," *Technology and Culture* 25, no.

4 (October 1984): 798–831. This emphasis can also be traced through articles in the Bureau's magazine, *Public Roads*, during the 1920s and 1930s.

11. See David Hounshell, "The Evolution of Industrial Research in the United States," in *Engines of Innovation: Industrial Research at the End of an Era*, ed. Richard S. Rosenbloom and William J. Spencer (Boston: Harvard Business School Press, 1996), 13–86. The contribution of science to the telephone, steel, oil, and electrical industries was vital but came after the engineering innovations that established these industries. See Billington and Billington, *Power Speed and Form*, 31–34, 50–53, 199–203, 210–222. On the ambition to be more scientific, see also Terry S. Reynolds, "Overview: The Engineer in 20th Century America," in *The Engineer in America: A Historical Anthology from Technology and Culture*, ed. Terry S. Reynolds (Chicago: University of Chicago Press, 1991), 169–190.

12. For the useful contributions of the Bureau's research, see *America's Highways 1776–1976*, 120–122; and Seely, *Building the American Highway System*, 71–87, 101–108. For early impact tests using vehicles, see Earl B. Smith, "The Motor Truck Impact Tests of the Bureau of Public Roads," *Public Roads* 3, no. 35 (March 1921): 3–36.

13. See again Seely, "The Scientific Mystique in Engineering," *Technology and Culture* 25, no. 4 (October 1984): 798–831. See also Leslie W. Teller, "Impact Tests on Concrete Pavement Slabs," *Public Roads* 5, no. 2 (April 1924): 1–14.

14. For the results of the Bates Test Road, see "Practical Lessons from the Bates Road Tests," *Engineering News-Record* 90, no. 2 (January 11, 1923): 57–61.

15. For the Pershing map, see *America's Highways 1776–1976*, 143; and for the new highway network, see Seely, *Building the American Highway System*, 54–65. For the 1921 Highway Act, see *United States Statutes at Large*, vol. 42 (1921–1923), Part 1, 212–219.

16. On the U.S. Highway numbering system, see *America's Highways 1776–1976*, 408. Some U.S. Highways later added more lanes in each direction. See also Dan McNichol, *The Roads That Built America: The Incredible Story of the U.S. Interstate System* (New York: Sterling Publishing, 2006), 67–74. For the mileage of surfaced roads, see the *Statistical Abstract of the United States for 1930*, 375, on which "high" type or modern roads refer to those paved with either kind of concrete, not to roads with surfaces of crushed stone that were merely treated or repaired with oil or asphalt.

17. For the social changes that occurred as automobile use spread, see David W. Jones, *Mass Motorization and Mass Transit: An American History and Policy Analysis* (Bloomington: Indiana University Press, 2008), 31–56. For an examination of the argument that streetcar service ended because auto and oil companies conspired to shut them down in the 1930s, see ibid., 64–68.

18. For the figure on road building as a proportion of federal work relief, see George H. Field et al., *Final Report on the WPA Program 1935–43* (Washington, DC: U.S. Government Printing Office, 1946), 47. For road-building after 1929, see Seely, *Building the American Highway System*, 88–99. For the increase in roadway mileage by 1940, see the *Historical Statistics of the United States: Colonial Times to 1970* (Washington, DC: Bureau of the Census, 1975), Part 2, 710.

19. On urban traffic and improvements to urban roads, see Seely, *Building the American Highway System*, 149–156; and Jones, *Mass Motorization and Mass Transit*, 68–85.

20. For the obsolescence of American highways in the 1930s, see *America's Highways 1776–1976*, 126–132. For improvements in oil refining and motor vehicles, see Billington and Billington, *Power Speed and Form*, 72–78, 203–205. The increasing engine power and speed of American automobiles can be followed in Beverly Rae Kimes, *Standard Catalogue of American Cars, 1805–1942*, 3rd ed. (Iola, WI: Krause Publications, 1996). MacDonald also observed the highways under construction in Nazi Germany in the 1930s that were designed for high speeds, and he believed that the United States would need similar highways. See Seely, *Building the American Highway System*, 147–148, 160–161.

21. For the calculation of stopping distance and super-elevation, see Charles M. Noble, "The Modern Express Highway," *ASCE Transactions* 102 (1937): 1068–1078.

22. For the urban orientation of MacDonald and the BPR, and criticisms of it, see Seely, *Building the American Highway System*, 156–164.

23. On the Merritt Parkway, see Bruce Radde, *The Merritt Parkway* (New Haven CT: Yale University Press, 1993). The parkway relieved congestion on U.S. Route 1, which ran alongside it to the south. For the controversy over toll roads, see Seely, *Building the American Highway System*, 165–177.

24. For the Bureau report, see *Toll Roads and Free Roads* (Washington, DC: U.S. Government Printing Office, 1939), published as 76th Congress, 1st Session, House Document No. 272.

25. For the serviceability of the nation's highways in World War II, see *America's Highways 1776–1976*, 144–145. For the postwar condition of American roads, see David P. Billington, "The Condition and Needs of the National Roadway Systems," BSE thesis (Princeton University, 1950).

26. For the new report on rural as well as urban highways, see *Interregional Highways* (Washington, DC: U.S. Government Printing Office, 1944), published as 78th Congress, 2nd Session, House Document No. 379.

27. For the 1944 Federal Highway Act, see *United States Statutes at Large*, vol. 58 (1944), Part 1, 838–843; and Seely, *Building the American Highway System*, 177–191. For the mileage of roads built from 1945 to 1950, see *Historical Statistics of the United States: Colonial Times to 1970*, Part 2, 710, and for automobile registrations, see ibid., 716.

28. For his recollections of the 1919 motor convoy, and his impression of German highways, see Dwight D. Eisenhower, *At Ease: Stories I Tell to Friends* (Garden City, NY: Doubleday, 1967), 157–167.

29. On MacDonald's last years and retirement, see Seely, *Building America's Highways*, 204–208; and Swift, *The Big Roads*, 157–161. The creation of the interstate system receives mention in Dwight D. Eisenhower, *The White House Years: Mandate for Change 1953–1956* (Garden City, NY: Doubleday, 1963), 547–549. See also Henry Moon, *The Interstate Highway System* (Washington, DC: Association of American Geographers, 1994). For the system's politics, construction, and

social impact, see Tom Lewis, *Divided Highways: Building the Interstate Highways, Transforming American Life* (New York: Viking, 1997). For the 1956 act, see *United States Statutes at Large,* vol. 70 (1956), Part 1, 374–402. Some bridges also collected tolls.

30. On the evolution of highway design standards, see *America's Highways 1776–1976*, 385–417. For highway design standards at the time the Interstate System began, see *A Policy on Design Standards: Interstate System, Primary System, Secondary and Feeder Roads* (Washington, DC: American Association of State Highway Officials, 1956). For the AASHO road testing program of 1958–1960, see Highway Research Board, *The AASHO Road Test* (Washington, DC: National Academy of Sciences–National Research Council, 1961–1962), 7 vols. The seven volumes are also titled Highway Research Board Special Reports 61A through 61G. For an overview and retrospective, see Kurt D. Smith, Kathryn A. Zimmermann, and Fred N. Finn, "The AASHO Road Test: Living Legacy for Highway Pavements," *TR News*, no. 232 (May–June 2004): 14–24. In 1973, the Association of State Highway Officials changed its name to the Association of State Highway and Transportation Officials (AASHTO).

31. For the numbering and signage of the Interstate System, see McNichol, *The Roads That Built America*, 118–123.

32. On the postwar growth of suburbs, see Robert Sobel, *The Great Boom, 1950–2000: How a Generation of Americans Created the World's Most Prosperous Society* (New York: St. Martin's Press, 2000). For the social role of the automobile, see Rudi Volti, *Cars and Culture: The Life Story of a Technology* (Baltimore, MD: Johns Hopkins University Press, 2006). For David Bohannon's work, starting in 1944, see "Big Dave Bohannon, Operative Builder by the California Method," *Fortune Magazine* 33, no. 4 (April 1946): 144–147, 190–200. On William Levitt's town on Long Island, New York, see Barbara M. Kelly, *Expanding the American Dream: Building and Rebuilding Levittown* (Albany: State University of New York Press, 1993); and for his town outside Philadelphia, see Richard Wagner and Amy Duckett Wagner, *Levittown* (Charleston, SC: Arcadia Publishing, 2010). See also *World War II and the American Dream: How Wartime Building Changed a Nation*, ed. Donald Albrecht (Cambridge, MA: MIT Press, 1995).

33. For the numbers of people and automobiles in 1950 and 1980, see the *Statistical Abstract of the United States for 1982–83*, 6, 615. For automobiles in the year 2000, see U.S. Bureau of the Census, *Statistical Abstract of the United States for 2010* (Washington, DC: U.S. Bureau of the Census, 2010), table 1034. For the mileage of paved roads from 1950 to 1995, see William H. Moore, ed., *National Transportation Statistics 2011* (Washington, DC: U.S. Department of Transportation, 2011), table 1-4. This report has no page numbers. There are no data in the report on road surfaces after 1995.

34. For the suburban and metropolitan population of the United States at the end of the twentieth century, see Frank Hobbs and Nicole Stoops, *Demographic Trends in the 20th Century*, U.S. Census Bureau, Census 2000 Special Reports, Series CENSR-4 (Washington, DC: U.S. Government Printing Office, 2002), 1, 32–33. For the shift of population and employment to the suburbs, see also Jones, *Mass Motorization and Mass Transit*, 108–136.

35. For the opposition to urban highways, see Swift, *The Big Roads*, 228–251, 264–313; and Raymond A. Mohl, "Stop the Road: Freeway Revolts in American Cities," *Journal of Urban History* 30, no. 5 (July 2004): 674–706. For the growth of suburbs and the consequences for America's cities and the suburbs themselves, see Kenneth T. Jackson, *Crabgrass Frontier: The Suburbanization of the United States* (New York: Oxford University Press, 1985); and Randall Bartlett, *The Crisis of America's Cities* (Armonk, NY: M. E. Sharpe, 1998). The federal government subsidized home ownership in the suburbs after World War II, as well as new highways, accelerating migration to suburbs that had begun earlier. Discrimination made it harder for racial minorities to join this migration. For the impact of motor vehicles on the railroads, see Stephen B. Goddard, *Getting There: The Epic Struggle between Road and Rail in the American Century* (Chicago: University of Chicago Press, 1994).

36. See Ralph Nader, *Unsafe at Any Speed: The Designed-In Dangers of the American Automobile* (New York: Grossman, 1965). The National Traffic and Motor Vehicle Safety Act passed in 1966 as Public Law 89-563. Before car traffic dominated them, city streets were often spaces shared by market stalls, slower-moving vehicles, recreational activity, and pedestrians. By the 1930s, city streets had become the preserve of motor vehicles; pedestrians could safely cross only at corners. For this change, see Peter Norton, *Fighting Traffic: The Dawn of the Motor Age in the American City* (Cambridge, MA: MIT Press, 2008).

37. For the challenges to the auto industry, see Jones, *Mass Motorization and Mass Transit*, 189–198; and *The Competitive Status of the U.S. Auto Industry* (Washington, DC: National Academy Press, 1982).

38. For bridge failures, see *Collapse of I-35W Highway Bridge, Minneapolis, Minnesota, August 1, 2007*, Accident Report NTSB/HAR-08-03 (Washington, DC: National Transportation Safety Board, 2008). The Mianus River Bridge in Connecticut had collapsed earlier, in 1983, when a standard connection holding the roadway deck in place failed. For an overview of how bridges fail, see Robert J. Dexter and John W. Fisher, "Fatigue and Fracture," in *Bridge Engineering Handbook*, ed. Wai-Fah Chen and Lian Duan (Boca Raton, FL: CRC Press, 2000), 53-1 to 53-23. For a report on bridges and other structures at risk, see the *2017 Infrastructure Report Card* (Reston, VA: American Society of Civil Engineers, 2017). See also David P. Billington, "One Bridge Does Not Fit All," *New York Times*, August 18, 2007, A13.

39. For Marilyn Jorgenson Reece (1926–2004), see the *Los Angeles Times*, May 21, 2004, B10. After graduating in 1948 from the University of Minnesota with a degree in civil engineering, Reece began her career with the California Division of Highways, now the California Department of Transportation (Caltrans). For an appreciation of the I-10/I-405 interchange, see Reyner Banham, *Los Angeles: The Architecture of Four Ecologies* (New York: Harper and Row, 1971), 88–90.

40. For the George Washington Bridge, see Billington and Billington, *Power Speed and Form*, 155–175; and for the Golden Gate Bridge, see David P. Billington, Jr., *Engineering the Golden Gate: The Interplay of Design and Experience* (San Francisco, CA: Golden Gate Bridge, Highway and Transportation District, 2016; public release forthcoming).

41. For the East Huntington Bridge, see Arvid Grant, "Design and Construction of the East Huntington Bridge," *PCI Journal* 32, no. 2 (January–February 1987): 20–29. The official name of the bridge is the Frank Gatski Memorial Bridge. The completed structure cost $38 million. The Swiss structural engineer Christian Menn designed the Leonard P. Zakim Memorial Bridge, a cable-stayed bridge that opened in Boston over the Charles River in 2003. This bridge cost about $105 million to construct.

42. For the development of tall buildings in Chicago, see David P. Billington, *The Tower and the Bridge: The New Art of Structural Engineering* (Princeton, NJ: Princeton University Press, 1985), 99–111. The first skyscraper was the Home Insurance Building, built in Chicago in 1884 and designed by William Le Baron Jenney. Curtain walls were already in use for smaller buildings.

43. On the Chrysler Building, see George C. Kingston, *William Van Alen, Fred T. Ley and the Chrysler Building* (Jefferson, NC: McFarland, 2017). For the Empire State Building, see *Building the Empire State*, ed. Carol Willis (New York: W. W. Norton and the Skyscraper Museum, 1998).

44. For the family and education of Fazlur Khan, see the biography by his daughter, Yasmin Sabina Khan, *Engineering Architecture: The Vision of Fazlur R. Khan* (New York: W. W. Norton, 2004), 25–39.

45. For Khan's insight of the framed tube, see Khan, *Engineering Architecture*, 85–92.

46. On Khan's design of the John Hancock Center, see Fazlur R. Khan, "John Hancock Center," *Civil Engineering* 37, no. 10 (October 1967): 38–42; and Fazlur R. Khan, "100-Storey John Hancock Center, Chicago: A Case Study of the Design Process," *Engineering Structures* 5, no. 1 (January 1983): 10–14. See also Khan, *Engineering Architecture*, 92–129; and Billington, *The Tower and the Bridge*, 234–242. The Chicago Hancock tower can also be termed a braced or trussed tube to distinguish it from a simpler framed tube without cross-braces. For a comparison of the steel amounts used in the John Hancock Center and the Empire State Building, see David P. Billington, "Technology and the Structuring of Cities," in *Small Comforts for Hard Times: Humanists on Public Policy*, ed. Michael Mahony and Florian Stuber (New York: Columbia University Press, 1977), 187–188. Converting the steel in the Hancock Center from pounds per square foot (psf) into kilogram-force per square meter (kgf/m^2) gives 145 kgf/m^2 for 29.7 psf and 206 kgf/m^2 for 42.2 psf.

47. See Billington, *The Tower and the Bridge*, 242. The Marina City apartments in Chicago, completed in 1964, were also a notable attempt to attract younger residents back to a city center. See Igor Marjanovic and Katerina Rüedi Ray, *Marina City: Bertrand Goldberg's Urban Vision* (New York: Princeton Architectural Press, 2010). For Fazlur Khan's place in structural engineering, see the papers in David P. Billington and Myron Goldsmith, eds., *Technique and Aesthetics in the Design of Tall Buildings* (Bethlehem, PA: Institute for the Study of the High-Rise Habitat, Lehigh University, 1986). Khan also designed in reinforced concrete; see again Khan, *Engineering Architecture*, 85–92. See also David P. Billington and Richard Alan Ellis, Jr., "Tall Concrete Structures: Ideas and Works of Fazlur Rahman Khan," *Concrete Journal* [Japan Concrete Institute] 40, no. 2 (2002): 29–34.

48. On Khan's collaboration with Bruce Graham, see Bruce J. Graham, "Collaboration in Practice Between Architect and Engineer," in *Technique and Aesthetics in the Design of Tall Buildings*, 1–3; and Billington, *The Tower and the Bridge*, 244–246.

49. For a critique of the machine-only perception of modern technology, see David P. Billington, "Structures and Machines: The Two Sides of Technology," *Soundings: An Interdisciplinary Journal* 58, no. 3 (fall 1974): 275–288. For an overview of the past, present, and future of the nation's civil engineering infrastructure, see Henry Petroski, *The Road Taken: America's Imperiled Infrastructure* (New York: Bloomsbury, 2016).

Chapter 4

1. For an overview of the physics of nuclear reactions, see Richard Wolfson, *Nuclear Choices: A Citizen's Guide to Nuclear Technology*, Alfred P. Sloan Foundation New Liberal Arts Series (Cambridge, MA: MIT Press, 1993), 1–118. For a more detailed presentation of nuclear energy that includes nuclear weapons and nuclear power, see David Bodansky, *Nuclear Energy: Principles, Practices, and Prospects* (Woodbury, NY: American Institute of Physics Press, 1996). For Einstein's life and work, see Albrecht Fölsing, *Albert Einstein: A Biography* (New York: Viking, 1997).

2. Letter from Albert Einstein to Franklin D. Roosevelt, August 2, 1939, Box 5, President's Safe File, Franklin D. Roosevelt Library and Museum, Hyde Park, New York. Einstein's letter to the President was drafted for him by the physicist Leo Szilard. On the Manhattan Project, see Henry DeWolf Smyth, *Atomic Energy for Military Purposes: The Official Report on the Development of the Atomic Bomb under the Auspices of the United States Government 1940–1945*, reprint ed. (Stanford, CA: Stanford University Press, 1989). See also Richard Rhodes, *The Making of the Atomic Bomb* (New York: Simon & Schuster, 1986).

3. On the 1942 Chicago test, see Dan Cooper, "Making It Possible," *Invention and Technology* 11, no. 1 (summer 1995): 10–21.

4. On the engineering of nuclear weapons, see Al Christman, "Making It Happen," *Invention and Technology* 11, no. 1 (summer 1995): 22–35.

5. For an overview of the AEC, see Alice L. Buck, *A History of the Atomic Energy Commission* (Washington, DC: U.S. Department of Energy, 1982), 1–10. For a more detailed account, see Richard G. Hewlett, Francis Duncan, and Oscar E. Anderson, Jr., *History of the United States Atomic Energy Commission* (Berkeley: University of California Press, 1989–1990), 3 vols.

6. For Rickover's life and career, see Francis Duncan, *Rickover: The Struggle for Excellence* (Annapolis, MD: Naval Institute Press, 2001). Duncan had access to classified papers and his book was vetted by the agencies concerned. For a more critical view of Rickover, see Norman Polmar and Thomas B. Allen, *Rickover* (New York: Simon & Schuster, 1981).

7. For Rickover's time at Oak Ridge and for the potential and challenge of engineering a nuclear submarine, see Duncan, *Rickover*, 93–100.

8. For Rickover's difficulties back in Washington, see Duncan, *Rickover*, 100–105. A former nuclear officer, Dr. Paul R. Schratz, in "Admiral Rickover and the Cult of Personality," *Air University Review* 34, no. 5 (July–August 1983): 96–101, argued that Rickover's personality was the likely cause of his postwar difficulties with others in the navy, not his religion or ethnicity. However, other biographers (cited above), while noting Rickover's capacity for abrasiveness, argued that he confronted serious anti-Semitic prejudice. Schratz noted that Rickover departed from the norm of the typical line officer, who rotated in and out of different assignments over the course of a career. Rickover's example of a lifetime career in one assignment clashed with the larger navy culture. Over time, though, nuclear officers came to be respected as an elite corps in their own right.

9. For the approval by Admiral Nimitz and Rickover's dual appointment with the AEC, see Duncan, *Rickover*, 105–109. For how his dual role worked, see Richard G. Hewlett and Francis Duncan, *Nuclear Navy 1946–1962* (Chicago: University of Chicago Press, 1974), 92. The dual appointment gave Rickover a degree of independence as well as influence in both organizations.

10. For the general design of the nuclear propulsion system, see L. H. Roddis, Jr., and J. W. Simpson, "The Nuclear Propulsion Plant of the USS Nautilus SSN-571," *Transactions of the Society of Naval Architects and Marine Engineers* 62 (1954): 491–521. Hereafter cited as *SNAME Transactions*.

11. On the two reactor prototypes, see Duncan, *Rickover*, 109–114. For the later cancellation of the sodium reactor, see ibid., 170.

12. For the 1946 paper, see A. M. Weinberg and F. H. Murray, "High Pressure Water as a Heat Transfer Medium in Nuclear Power Plants," April 10, 1946, reprinted as an appendix to chapter 3 in Alvin M. Weinberg, *The First Nuclear Era: The Life and Times of a Technological Fixer* (Woodbury, NY: American Institute of Physics Press, 1994), 72–75. The co-author's full name was Forrest H. Murray. See also ibid., 41–43, 49–51, and 58–61.

13. The Rickover quote is from Theodore Rockwell, *The Rickover Effect: The Inside Story of How Admiral Hyman Rickover Built the Nuclear Navy* (New York: John Wiley and Sons, 1992), 87.

14. For the use of zirconium and hafnium, see Rockwell, *The Rickover Effect*, 87–90, 131–133; and Hewlett and Duncan, *Nuclear Navy*, 59, 139–142. The reactor would also need a start-up source of neutrons, which came from radium and beryllium. For the supply of these, see Paul Litt, *Isotopes and Innovation: MDS Nordion's First Fifty Years, 1946–1996* (Ottawa, ON: MDS Nordion, 2000), 125. Once fission began, the neutrons in the uranium fuel sustained the chain reaction.

15. Quotes are from Rockwell, *The Rickover Effect*, 92.

16. For the building and testing of the Mark I, see L. H. Roddis, Jr., and J. W. Simpson, "The Nuclear Propulsion Plant of the USS Nautilus SSN-571," *SNAME Transactions* 62 (1954): 491–521; and Hewlett and Duncan, *Nuclear Navy*, 164–168. For the shipbuilding contract, see Rockwell, *The Rickover Effect*, 96–99.

17. On Rickover's struggle for promotion, see Duncan, *Rickover*, 116–131. To support him, the *Time* correspondent, Clay Blair, subsequently published an account of Rickover and his efforts, with the private assistance of Rickover and his wife. See Clay Blair, Jr., *The Atomic Submarine and*

Admiral Rickover (New York: Henry Holt, 1954). For the assistance of the Rickovers to Blair, see Duncan, *Rickover*, 134–135.

18. The *Nautilus* was the fourth U.S. warship to bear the name and the second modern submarine to do so, the first submarine *Nautilus* having served with distinction in World War II. See the *Dictionary of American Naval Fighting Ships*, 8 vols. (Washington, DC: U.S. Government Printing Office, 1959–1981), vol. 5 (1970), 26–29. In L. H. Roddis, Jr., and J. W. Simpson, "The Nuclear Propulsion Plant of the USS Nautilus SSN-571," *SNAME Transactions* 62 (1954): 497, the ship is compared to Jules Verne's fictional predecessor. *Jane's Fighting Ships 1955–56*, ed. Raymond V. B. Blackman (New York: McGraw-Hill, 1956), 425, gives a length and width to the nuclear ship of 300 × 28 feet, a crew complement of 12 officers and 89 men, and a displacement of 2,980 tons for the ship and 3,180 tons fully loaded. The entry for the nuclear *Nautilus* (SSN-571) in the *Dictionary of American Naval Fighting Ships*, vol. 5 (1970), 28–29, gives the weight submerged as 4,000 tons and the length and width as 319 × 27.8 feet, with a crew complement of 105. Both sources agree that the speed submerged was greater than 23 knots (i.e., greater than 26.4 mph, or 42.4 km/hr). This was more than twice the submerged speed of a World War II diesel–electric submarine.

19. Quote is from "Statement of Admiral H. G. Rickover, USN, Director, Naval Nuclear Propulsion Program, before the Subcommittee on Energy Research and Production of the Committee on Science and Technology, United States House of Representatives," May 24, 1979, p. 105. Copy attached to the testimony of Admiral Skip Bowman, director of the naval nuclear program when the space shuttle *Columbia* failed in 2003, on how the navy managed safety. For Admiral Bowman's testimony, see *NASA's Organizational and Management Challenges in the Wake of the Columbia Disaster*, Hearing before the Committee on Science, House of Representatives, 108th Congress, 1st Session, October 29, 2003 (Washington, DC: U.S. Government Printing Office, 2004), 17–24.

20. On the selection of the first captain of the *Nautilus*, Commander Eugene P. Wilkinson, see Duncan, *Rickover*, 132–134. For the completion and sea trials of the submarine, see ibid., 135–140. For the voyage under the North Pole, see Rockwell, *The Rickover Effect*, 247–249. The Arctic voyages of the USS *Nautilus* and the USS *Skate* are in the respective articles on each ship in the *Dictionary of American Naval Fighting Ships*. The *Skate* is listed as *Skate II*.

21. On the *Thresher* loss and the navy's response, see Duncan, *Rickover*, 192–197.

22. On Rickover's later years, see Duncan, *Rickover*, 151–305. For his relationship with President Carter, see ibid., 257–276. For the President's recollections of Rickover, see Jimmy Carter, *Why Not the Best?* (Nashville, TN: Broadman Press, 1975), 53–58. Quote is from p. 55.

23. For a technical overview of civilian nuclear energy, see Wolfson, *Nuclear Choices*, 121–181. For the early history of civilian nuclear power, see George T. Mazuzan and J. Samuel Walker, *Controlling the Atom: The Beginnings of Nuclear Regulation, 1946–1962* (Berkeley: University of California Press, 1985); and J. Samuel Walker, *Containing the Atom: Nuclear Regulation in a Changing Environment, 1963–1971* (Berkeley: University of California Press, 1992). For the Atomic Energy Act of 1954 (Public Law 83-703) and the Price–Anderson Act (Public Law 85-256), see *United States Statutes at Large*, vol. 68 (1954), 919–961; and vol. 71 (1957), 576–579.

24. For the Shippingport nuclear power station, see William Beaver, *Nuclear Power Goes On-Line: A History of Shippingport* (Westport, CT: Greenwood Press, 1990).

25. For the electricity generated in the United States from nuclear energy in 1960 and 1978, see the *Statistical Abstract of the United States for 1980*, vol. 611 (table 1043). For the operating and construction costs of nuclear plants, see Bodansky, *Nuclear Energy*, 8–10, 307–310. For the National Environmental Policy Act of 1969 (Public Law 91-190), see *United States Statutes at Large*, vol. 83 (1969), 852–856. The volume of nuclear waste could be reduced through reprocessing but to do so would create more of an easily weaponized material, plutonium, and the U.S. government has not wanted to produce more plutonium.

26. On the accident at Three Mile Island, see J. Samuel Walker, *Three Mile Island: A Nuclear Crisis in Historical Perspective* (Berkeley: University of California Press, 2004).

27. See *Report of the President's Commission on the Accident at Three Mile Island—The Need for Change: The Legacy of TMI* (Washington, DC: President's Commission on the Accident at Three Mile Island, 1979). For the actions taken after the crisis, see Walker, *Three Mile Island*, 209–225.

28. For a list of plant closures, see Bodansky, *Nuclear Energy*, 13. For the escalation of costs and a comparison of costs between plants, see Bernard L. Cohen, *The Nuclear Energy Option: An Alternative for the 90s* (New York: Plenum Press, 1990), 145–157. Cohen attributed most of the increased cost of nuclear plant construction after Three Mile Island to the need to change designs and equipment and hire more expensive labor, often after significant construction had been done, to meet new regulatory standards. But he also documented (on p. 148) the difference in cost between the "median" nuclear plant and the "best" nuclear plant and he compared these with coal-burning plants. The "median" nuclear plant cost twice as much to build in the 1980s as the "best" nuclear one, and coal plants were cheaper than both, suggesting that nuclear power management at the time still needed to improve.

29. For the nuclear share of net electrical generating capacity in the United States since 1990, see the *Statistical Abstract of the United States for 2011*, 593. For the increased output of nuclear plants, see Bodansky, *Nuclear Energy*, 11–12.

30. For an overview of nuclear fusion research, see Garry McCracken and Peter Stott, *Fusion: The Energy of the Universe* (Amsterdam: Elsevier Academic Press, 2005).

31. For the development and frustrations of fusion research in the twentieth century, see Robin Herman, *Fusion: The Search for Endless Energy* (Cambridge: Cambridge University Press, 1990). For a more critical view, see Charles Seife, *Sun in a Bottle: The Strange History of Fusion and the Science of Wishful Thinking* (New York: Viking Press, 2008). For the future of fusion research using lasers, see National Research Council, *An Assessment of the Prospects for Inertial Fusion Energy* (Washington, DC: National Academies Press, 2013). For the future of magnetic fusion, see National Academies of Science, Engineering, and Medicine, *Final Report of the Committee on a Strategic Plan for U.S. Burning Plasma Research* (Washington, DC: National Academies Press, 2019). These assessments argue that a modest continued investment is in the national interest despite the uncertain timeframe for success.

32. For the Gulf of Mexico oil spill in 2010, see *Deep Water: The Gulf Oil Disaster and the Future of Offshore Drilling: Report to the President of the National Commission on the BP Deepwater Horizon Oil Spill and Offshore Drilling* (Washington, DC: U.S. Government Printing Office, 2011); and *Report of Investigation into the Circumstances Surrounding the Explosion, Fire, Sinking and Loss of Eleven Crew Members Aboard the Mobile Drilling Unit Deepwater Horizon in the Gulf of Mexico April 20–22, 2010* (Washington, DC: U.S. Coast Guard, 2013).

33. A report from the National Research Council, *Induced Seismicity Potential in Energy Technologies* (Washington, DC: National Academies Press, 2013), notes that the disposal of wastewater into used injection wells is more likely the cause of seismic disturbance than fracking itself, but since disturbances can continue long after the drilling has ended, the seriousness of the danger is clear.

34. For nuclear weapons and their proliferation, see Wolfson, *Nuclear Choices*, 289–454.

35. For the arguments on both sides, see Michio Kaku and Jennifer Trainer, *Nuclear Power, Both Sides: The Best Arguments for and Against the Most Controversial Technology* (New York: W. W. Norton, 1982). For an assessment of civilian nuclear power, see MIT Energy Initiative, *The Future of Nuclear Power: An Interdisciplinary Study* (Cambridge, MA: Massachusetts Institute of Technology, 2003).

Chapter 5

1. For aircraft engine development during the interwar period, see Edward W. Constant II, *The Origins of the Turbojet Revolution* (Baltimore, MD: Johns Hopkins University Press, 1980), 117–150; for an overview of aircraft propulsion, see ibid., 247–265. For jet propulsion, see also John D. Anderson, Jr., *Introduction to Flight*, 3rd ed. (New York: McGraw-Hill, 1989), 491–501. At the end of the nineteenth century, the branch of science nearest to aerodynamics, fluid dynamics, had no influence on the Wright brothers. See John D. Anderson, Jr., *A History of Aerodynamics and Its Impact on Flying Machines* (Cambridge: Cambridge University Press, 1997), 242–243. For advances in aerodynamic theory after the historic 1903 flight of the Wrights, see ibid., 244–369. For changes to aircraft design, see also David P. Billington and David P. Billington, Jr., *Power, Speed, and Form: Engineers and the Making of the Twentieth Century* (Princeton, NJ: Princeton University Press, 2006), 126, 199–200, 206–219. For the improvements to gasoline, see ibid., 73–77.

2. For the sharp increase in air resistance, or aerodynamic drag, on the propellers at propeller speeds approaching the speed of sound, see Anderson, *A History of Aerodynamics*, 382–400. The speed of sound is 761 mph, or 1,224 km/hr, at sea level with a temperature of 68°F (20°C). At a cruising altitude of 10,000 feet (3,048 m) at 23°F (–4.8 C), the speed of sound is 734 mph (1,182 km/hr).

3. For the performance of the Douglas DC-3, see *Jane's All the World's Aircraft 1938*, ed. C. G. Grey and Leonard Bridgman (London: Sampson Low, Marston, 1938), 264–265.

4. On the use of turbines in the textile industry, see David P. Billington, *The Innovators: The Engineering Pioneers Who Made America Modern* (New York: John Wiley and Sons, 1996), 79–94. For

their use in electric power generation, see chapter 2 in the present book. For the work of Charles A. Parsons (1854–1931), see Constant, *The Origins of the Turbojet Revolution*, 69–77.

5. For early attempts to make an internal combustion engine with a turbine, see Constant, *The Origins of the Turbojet Revolution*, 89–98. For the work of Sanford Moss (1872–1946), see ibid., 94–96, 98. The hot gases from burning the fuel powered his turbine. Moss later made important contributions to aircraft engine design. See his obituary in the *New York Times*, November 11, 1946, 27, col. 1.

6. For the development of mechanical air compressors and supercharging, see Constant, *The Origins of the Turbojet Revolution*, 83–89, 122–125.

7. For Alan Arnold Griffith (1893–1963) and his proposal, which went unpublished, see Constant, *The Origins of the Turbojet Revolution*, 110–114. The idea was later adopted in aviation as the "turboprop" engine.

8. See Edgar Buckingham, "Jet Propulsion for Airplanes," Report No. 159, in the *Ninth Annual Report of the National Advisory Committee on Aeronautics* (Washington, DC: U.S. Government Printing Office, 1924), 75–90. For the resistance to jet propulsion in the 1920s and 1930s, see James R. Hansen, *Engineer in Charge: A History of the Langley Aeronautical Laboratory, 1917–1958* (Washington, DC: National Aeronautics and Space Administration, 1987), 219–226.

9. For the committee report, see U.S. Navy, Bureau of Ships, *An Investigation of the Properties of the Gas Turbine for Marine Propulsion*, Technical Bulletin, No. 2 (January 1941). Copy in the National Academy of Sciences Archives (NAS-NRC Archives: Central File: ADM: ORG: Committee on Gas Turbines: Investigation of Possibilities of Gas Turbine for Marine Propulsion: Report to Secretary of Navy: 1940–1941). The report was submitted to the Secretary of the Navy on June 10, 1940, and the comments on aircraft propulsion are in ibid., 37. See also Theodore von Kármán with Lee Edson, *The Wind and Beyond: Theodore von Kármán Pioneer in Aviation and Pathfinder in Space* (Boston: Little, Brown, 1967), 225. Von Kármán served on the committee but stated in his memoir that he signed the report without having read it.

10. For Whittle's early life and service as an RAF apprentice and cadet, see John Golley with Sir Frank Whittle, *Whittle: The True Story* (Washington, DC: Smithsonian Institution Press, 1987), 1–26. For a more recent account, see Andrew Nahum, *Frank Whittle and the Invention of the Jet* (Cambridge: Icon Books, 2004).

11. See Sir Frank Whittle, "The Birth of the Jet Engine in Britain," in *The Jet Age: Forty Years of Jet Aviation*, ed. Walter J. Boyne and Donald S. Lopez (Washington, DC: National Air and Space Museum/Smithsonian Institution, 1979), 3–25. In ibid., 3, Whittle briefly mentions his thesis. He made a case for the turbine at high altitudes in Flight Cadet F.W., "Speculation," *Royal Air Force Cadet College Magazine* 8, no. 2 (autumn 1928): 106–110.

12. Quote is from Sir Frank Whittle, *Jet, The Story of a Pioneer* (London: Frederick Muller, 1953), 24–25.

13. For his patent, see F. Whittle, "Improvements Relating to the Propulsion of Aircraft and Other Vehicles," Patent Specification No. 347,206, applied January 16, 1930, accepted April 16, 1931 ([London]: His Majesty's Stationery Office, 1931). See also Air Commodore F. Whittle, "The Early History of the Whittle Jet Propulsion Gas Turbine," James Clayton Memorial Lecture for 1945 (London: Institution of Mechanical Engineers, 1945), 419–435, esp. 419–420.

14. For Whittle's inability to interest higher authority in his jet engine, see Whittle, "The Birth of the Jet Engine in Britain," 4. The £5 he needed in 1935 would have been worth about £270 (or $421) in 2010.

15. On the formation of Power Jets in 1936, see Whittle, "The Birth of the Jet Engine in Britain," 6. One of the officers, R. D. Williams, was a Cranwell cadet whom Whittle had known, and the other was a friend of Williams, J. C. B. Tinling, who helped connect Whittle to the investment bank of Falk and Partners. The scientist and philosopher Lancelot Law Whyte, one of the Falk partners, led the funding for the jet engine.

16. For the performance goals of Whittle's first jet engine and airplane, see Air Commodore F. Whittle, "The Early History of the Whittle Jet Propulsion Gas Turbine," James Clayton Memorial Lecture 1945, 423.

17. For the development of the Whittle engine to 1939, see Whittle, "The Birth of the Jet Engine in Britain," 5–12.

18. See again Whittle, "The Birth of the Jet Engine in Britain," 12–15.

19. For the Gloster E.28/39 and the Gloster Meteor, see *Jane's All the World's Aircraft 1945–46*, ed. Leonard Bridgman (New York: McGraw-Hill, 1946), 39c–40c. The entries give little technical data. An entry on the Power Jets W1 and W2 engines, however, gives weight, fuel consumption, and thrust. See ibid., 6d–7d. On the wartime U.S. jet engines and airplanes, see ibid., 204c–205c (Bell Airacomet) and 271c–272c (Lockheed P-80). Whittle retired from the RAF in 1948.

20. Although his interest in jet propulsion began earlier, Ohain studied Whittle's patents in the 1930s before making a test engine design of his own. For his and other German work on jet engines, see Hans von Ohain, "The Evolution and Future of Aeropropulsion Systems," in *The Jet Age*, 25–46; and Constant, *The Origins of the Turbojet Revolution*, 194–213. For the development and deployment of jet aircraft by Nazi Germany, see ibid., 230–234. See also *Jane's All the World's Aircraft 1945–46*, 120c–122c, 132c, and 141c–142c.

21. Alan Griffith's turboprop engine employed axial flow, and he envisioned an axial-flow jet engine once Whittle had proved that a jet engine was possible. See Constant, *The Origins of the Turbojet Revolution*, 214–215. For the change to axial flow, see Brian John Nichelson, "Early Jet Engines and the Transition from Centrifugal to Axial Compressors: A Case Study in Technological Change," PhD dissertation (University of Minnesota, 1988).

22. The first American military jet airplane with swept-back wings, the F-86 Sabre, used an axial-flow compressor in its engine. See *Jane's All the World's Aircraft 1953–54*, ed. Leonard Bridgman (New York: McGraw-Hill, 1953), 257–259. On the evolution of swept-back wings, see Anderson, *A*

History of Aerodynamics, 423–430. The idea began with a 1935 proposal by a German aerodynamic theorist, Adolf Busemann.

23. For the Comet, see *Jane's All the World's Aircraft 1953–54*, 62–65; and Timothy Walker and Scott Henderson, *The First Jet Airliner: The Story of the de Havilland Comet* (Newcastle-upon-Tyne: Scoval Publishing, 2000). See *Jane's All the World's Aircraft 1960–61*, ed. John W. R. Taylor (New York: McGraw-Hill, 1960), 32–34, for the redesigned Comet, and ibid., 267–269, for the Boeing 707.

24. See *Jane's All the World's Aircraft 1960–61*, 302–304, for the Douglas DC-8. For the changes in rail and air travel, see the *Statistical Abstract of the United States 1982–83*, 103rd ed. (Washington, DC: U.S. Government Printing Office, 1982), 607 (table 1044). A passenger-mile is a measure of one person moving one mile. For the decline in passenger rail travel, see also John F. Stover, *The Life and Decline of the American Railroad* (New York: Oxford University Press, 1970), 223–233; and Federal Highway Administration, *America's Highways 1776–1976: A History of the Federal-Aid Program* (Washington, DC: U.S. Federal Highway Administration, 1977), 117.

25. For the turbofan engine and its fuel economy, see Anderson, *Introduction to Flight*, 499–500. Kerosene replaced gasoline as the fuel for jet aircraft.

26. For the Boeing 707, see again *Jane's All the World's Aircraft 1960–61*, 267–269. For the Boeing 747, see *Jane's All the World's Aircraft 1972–73*, ed. John W. R. Taylor (New York: McGraw-Hill, 1972), 274–276.

27. For the air traffic control system in the United States, see Michael S. Nolan, *Fundamentals of Air Traffic Control*, 2nd ed. (Belmont, MA: Wadsworth, 1994). For the growth of airlines, see T. A. Heppenheimer, *Turbulent Skies: The History of Commercial Aviation* (New York: John Wiley and Sons, 1995); and for the growth of airports, see Alastair Gordon, *Naked Airport* (New York: Henry Holt, 2004).

28. For Vaughn's proposal and its fate, see Hansen, *Engineer in Charge*, 225–226.

29. For his seminal 1903 article, "A Rocket into Cosmic Space," see K. E. Tsiolkovskiy, *Works on Rocket Technology*, NASA Technical Translation (TT) F-243 (Washington, DC: National Aeronautics and Space Administration, 1965), 24–59. For Tsiolkovsky's contribution to spaceflight, see Asif A. Siddiqi, *The Red Rocket's Glare: Spaceflight and the Soviet Imagination, 1857–1957* (Cambridge: Cambridge University Press, 2010), 16–30.

30. The authorized biography of Robert Hutchings Goddard is Milton Lehman, *This High Man: The Life of Robert H. Goddard* (New York: Farrar, Straus, 1963). For a recent and more critical study, see David A. Clary, *Rocket Man: Robert H. Goddard and the Birth of the Space Age* (New York: Theia, 2003).

31. Multiple-stage rockets first appeared in medieval China. For his early findings, see Robert H. Goddard, "A Method of Reaching Extreme Altitudes," *Smithsonian Miscellaneous Collections* 71, no. 2 (1919): 69 pp. with plates. For a summary, see Robert H. Goddard, "A Method of Reaching Extreme Altitudes," *Nature* 105, no. 2652 (August 26, 1920): 809–811. For an editorial doubting

the possibility of travel in a vacuum, see the *New York Times*, January 13, 1920, 12, col. 5. For public reaction and Goddard's response, see Clary, *Rocket Man*, 87–97. The *New York Times* printed a retraction of its 1920 editorial on July 17, 1969, 43, cols. 6–7.

32. For Goddard's 1926 rocket launch in Massachusetts, see *The Papers of Robert Goddard*, ed. Esther C. Goddard and G. Edward Pendray (New York: McGraw-Hill, 1970), 3 vols., 2: 588–589. Goddard's 1926 rocket motor received liquid oxygen and gasoline from cylinders attached under the motor to fuel the ignition. He believed that carrying the fuel below would keep the motor vertical. This proved untrue and his later rockets carried liquid oxygen and gasoline above the combustion chamber. For the 1929 launch, see Robert H. Goddard, *Rocket Development: Liquid-Fuel Rocket Research 1929–1941* (Englewood Cliffs, NJ: Prentice-Hall, 1948; 1960), xvii–xviii; Lehman, *This High Man*, 152–156; and Clary, *Rocket Man*, 119–122, 133–136.

33. For Lindbergh's interest, see the preface he wrote to Lehman, *This High Man*, xiii–xv. For the support of Goddard's work by the Daniel Guggenheim Fund for the Promotion of Aeronautics, see Richard Hallion, *Legacy of Flight: The Guggenheim Contribution to American Aviation* (Seattle: University of Washington Press, 1977), 174–177.

34. For Goddard's work at Roswell from 1930 to 1935, see Robert H. Goddard, "Liquid-Propellant Rocket Development," *Smithsonian Miscellaneous Collections* 95, no. 3 (1936–1937): 10 pp. with plates. For his 1937 launch, see Goddard, *Rocket Development*, 106–108.

35. For the work at Caltech, see von Kármán, *The Wind and Beyond*, 234–241, 245; and Hallion, *Legacy of Flight*, 177–204.

36. See Hermann Oberth, *Die Rakete zu den Planetenräumen* (Munich: R. Oldenbourg, 1923) and *Wege zur Raumschiffahrt* (Berlin: R. Oldenbourg, 1929). For an authorized account of his life, see Helen B. Walters, *Hermann Oberth: Father of Space Travel* (New York: Macmillan, 1962).

37. On the German rocket program, see Norman Longmate, *Hitler's Rockets: The Story of the V-2s* (London: Hutchinson, 1985). For a critical view of von Braun's activity in World War II, see Wayne Biddle, *Dark Side of the Moon: Wernher von Braun, The Third Reich, and the Space Race* (New York: W. W. Norton, 2009), 138–141.

38. For engineering data on the V-1 missile and V-2 rocket, see *Jane's All the World's Aircraft, 1945–46*, 147c–149c. The Germans gave the V-2 rocket the technical designation A-4.

39. See Charles E. Yeager, "Flying Jet Aircraft and the Bell XS-1," in *The Jet Age*, 101–109. The U.S. Air Force organized as a separate service in 1947. Rocket planes, such as the North American X-15, which began flights in 1959, lifted pilots above 50 miles (80 km) in altitude. The U.S. Air Force defined this altitude as the boundary of outer space.

40. For the creation of NASA, see David F. Portree, *NASA's Origins and the Dawn of the Space Age*, Monographs in Aerospace History No. 10 (Washington, DC: NASA History Office, 1998). On the development of the Minuteman missile, see J. D. Hunley, "Minuteman and the Development of Solid-Rocket Launch Technology," in *To Reach the High Frontier: A History of U.S. Launch Vehicles*, ed. Roger D. Launius and Dennis R. Jenkins (Lexington: University Press of Kentucky, 2002),

229–300. For the Polaris missile, see Graham Spinardi, *From Polaris to Trident: The Development of U.S. Fleet Ballistic Missile Technology* (Cambridge: Cambridge University Press, 1994).

41. On the Soviet space program at the time, see Asif Siddiqi, *Challenge to Apollo: The Soviet Union and the Space Race, 1945–1974*, NASA SP-2000-4408 (Washington, DC: NASA, 2000).

42. For the decision to go to the Moon, see John M. Logsdon, "The Evolution of U.S. Space Policy and Plans," in *Exploring the Unknown: Selected Documents in the U.S. Civil Space Program*, NASA SP-4407, 6 vols., ed. John M. Logsdon (Washington, DC: NASA, 1995), 1: 377–393. For President Kennedy's call to Congress, see ibid., 453.

43. For the work of Draper and the MIT laboratory, see Robert A. Duffy, *Charles Stark Draper: A Biographical Memoir* (Washington, DC: National Academy of Sciences, 1994); and Eldon C. Hall, *Journey to the Moon: A History of the Apollo Guidance Computer* (Reston, VA: Institute of Aeronautics and Astronautics, 1996). See also David Mindell, *Digital Apollo: Human and Machine in Six Lunar Landings* (Cambridge, MA: MIT Press, 2008).

44. For the work of the NASA Marshall Space Flight Center, see Andrew J. Dunar and Stephen P. Waring, *Power to Explore: A History of Marshall Space Flight Center, 1960–1990*, NASA SP-4313 (Washington, DC: NASA, 1999).

45. For the debate over how to travel to and from the Moon, see James S. Hansen, *Enchanted Rendezvous: John C. Houbolt and the Genesis of the Lunar-Orbit Rendezvous Concept*, Monographs in Aerospace History, No. 4 (Washington, DC: NASA, 1995).

46. For the construction of the launch complex at Cape Canaveral, see Charles D. Benson and William Barnaby Faherty, *Moonport: A History of Apollo Launch Facilities and Operations*, NASA SP-4204 (Washington, DC: NASA, 1978). The author's father, David P. Billington, was project engineer for the control center and service tower at Cape Canaveral Launch Complex 36. See Major Warren Daily and Anton Tedesko, "Take-Off Point for a Trip into Outer Space," *Engineering News-Record* 165, no. 19 (November 10, 1960): 38–39. LC 36 was the launch site for the Pioneer missions to the outer planets. The Apollo missions launched from LC 39.

47. For the NASA accounts of the Mercury and Gemini programs, see Loyd S. Swenson, Jr., James M. Grimwood, and Charles C. Alexander, *This New Ocean: A History of Project Mercury*, NASA SP-4201 (Washington, DC: NASA, 1966; reprint ed., 1999); and Barton C. Hacker and James M. Grimwood, *On the Shoulders of Titans: A History of Project Gemini*, NASA SP-4203 (Washington, DC: NASA, 1977; reprint ed., 2002). The literature on the Apollo project is more extensive, but a starting point is Courtney G. Brooks, James M. Grimwood, and Loyd S. Swenson, Jr., *Chariots for Apollo: A History of Manned Lunar Spacecraft*, NASA SP-4205 (Washington, DC: NASA, 1979; reprint ed., 2009); and *Apollo: Expeditions to the Moon*, ed. Edward M. Cortright (Washington, DC: NASA, 1975; reprint ed. 2009).

48. For the thrust of the Saturn engines, see the *Saturn V Flight Manual: SA-507* (Huntsville, AL: NASA George C. Marshall Space Flight Center, 1969), 2-3 and 2-14. The thrust figure of 1.5 million pounds is at sea level. For the origin and development of the Saturn rocket, see Roger E. Bilstein, *Stages to Saturn: A Technological History of the Apollo/Saturn Launch Vehicles* (Gainesville:

University Press of Florida, 2003). Bilstein lists the major contractors and subcontractors for the Saturn rocket in ibid., 424–438.

49. For the first-stage rocket and the difficulties with combustion, see Bilstein, *Stages to Saturn*, 99–101, 109–116.

50. For the challenges of insulating the second stage rocket, see Bilstein, *Stages to Saturn*, 214–215. These tanks were thinner on the second stage than on the third, which also used both liquid oxygen and liquid hydrogen. External insulation was a better way to add material strength to the tanks.

51. For the design of the Saturn V and Apollo spacecraft, see *Apollo Summary Program Report*, JSC-09423 (Houston, TX: NASA Lyndon B. Johnson Space Flight Center, 1975), Section 4. For the Vehicle Assembly Building, see the report of its designer, Anton Tedesko, "Design of the Vertical Assembly Building," *Civil Engineering* 35, no. 1 (January 1965): 45–49. In 1965 the structure was renamed the Vehicle Assembly Building. North American Aviation also built the Command and Service Modules. Grumman Aerospace built the Lunar Module. North American later became part of Boeing and Grumman merged to form Northrop Grumman.

52. For the NASA report on the tragedy, see *Apollo 204 Review Board: Report to the Administrator, National Aeronautics and Space Administration* (Washington, DC: U.S. Government Printing Office, 1967). To honor the astronauts who lost their lives, the mission was redesignated Apollo 1 from Apollo-Saturn 204. Both houses of Congress also held hearings on the tragedy. For the Senate report, see 90th Congress, 2nd Session, Senate Report No. 956, *Apollo 204 Accident, Report of the Committee on Aeronautical and Space Sciences, United States Senate, with Additional Views* (Washington, DC: U.S. Government Printing Office, 1968).

53. For the Apollo flights, see the *Apollo Summary Program Report*, Sections 2.3 and 2.4.

54. For a first-hand account of the Apollo 8 mission, see "Our Moon Journey: Frank Borman, James Lovell, and Bill Anders Write Their Own Accounts of the Apollo 8 Flight," *Life Magazine* 66, no. 2 (January 17, 1969): 26–31. The night before their departure, Charles Lindbergh visited the Apollo 8 crew.

55. For the Apollo 11 mission, see the *Apollo 11 Mission Report*, NASA SP-238 (Washington, DC: NASA, 1971). For a transcript of the flight, see National Aeronautics and Space Administration, *Apollo 11 Technical Air-to-Ground Voice Transcription* (Houston, TX: Manned Spacecraft Center, 1969). For first-hand accounts, see Edwin E. "Buzz" Aldrin, Jr., with Wayne Warga, *Return to Earth* (New York: Random House, 1973); and Michael Collins, *Carrying the Fire: An Astronaut's Journeys* (New York: Farrar, Straus and Giroux, 1974). Neil Armstrong did not write an account for a wider public but authorized a biography that included one. See James R. Hansen, *First Man: The Life of Neil Armstrong* (New York: Simon & Schuster, 2005).

56. See Roger D. Launius, *Apollo: A Retrospective Analysis*, Monographs in Aerospace History, No. 3, NASA SP-2004-4503 (Washington, DC: NASA, 1994; reprint, 2004). See also W. D. Kay, *Defining NASA: The Historical Debate over the Agency's Mission* (Albany: State University of New York Press, 2005). For the benefits and drawbacks of scaling down NASA's budget, see Howard

E. McCurdy, *Faster Better Cheaper: Low-Cost Innovation in the U.S. Space Program* (Baltimore, MD: Johns Hopkins University Press, 2001).

57. For Skylab, see *Skylab: Our Nation's First Space Station*, ed. Leland F. Belew (Washington, DC: NASA, 1977); and for the Space Shuttle, see T. A. Heppenheimer, *The Space Shuttle Decision: NASA's Search for a Reusable Space Vehicle* (Washington, DC: NASA, 1999). For the specifications of the shuttle, formally designated the Space Transportation System (STS), see *Jane's All the World's Aircraft 1984–85* (London: Jane's Publishing, 1984), 799–801. The two shuttle rocket boosters at liftoff delivered 7,725,000 pounds of thrust.

58. For the U.S. astronaut corps, see *Information Summaries: Astronaut Fact Book*, NP-2013-04-003-JSC (Washington, DC: NASA, 2013).

59. For an overview of the International Space Station, see *Engineering Challenges to the Long-Term Operation of the International Space Station* (Washington, DC: National Academy Press, 2000). With the cooperation of NASA, the private U.S. launch firm SpaceX began sending American astronauts to the space station in 2020.

60. For the statistics on the 1960s space program, see Jane Van Nimmen and Leonard C. Bruno with Robert L. Rosholt, *NASA Historical Data Book, 1958–1968*, NASA SP-4012 (Washington, DC: NASA, 1976), 6 vols., vol. 1 NASA Resources, 5 (personnel), 6 (budget), and 8 (Saturn portion of the budget).

61. For the management of NASA in the 1960s, see W. Henry Lambright, *Powering Apollo: James E. Webb of NASA* (Baltimore, MD: Johns Hopkins University Press, 1998).

62. For the balance between NASA supervision and private involvement, see Launius, *Apollo: A Retrospective Analysis*, 7–10.

63. Individual insight was still important at a more detailed level. One of the many smaller innovations in the space program that resulted from the insight of one or two individuals was the counterintuitive discovery that the heat on a space capsule reentering the atmosphere was inversely proportional to the drag of the capsule. For this reason, the heat shields on Apollo Command Modules were broadly rounded instead of pointed. See H. Julian Allen and A. J. Eggers, Jr., *A Study of the Motion and Aerodynamic Heating of Ballistic Missiles Entering the Earth's Atmosphere at High Supersonic Speeds*, NACA Technical Report No. 1381 (Washington, DC: National Advisory Committee on Aeronautics, 1958).

64. See Gene Kranz, *Failure Is Not an Option: Mission Control from Project Mercury to Apollo 13 and Beyond* (New York: Simon & Schuster, 2009).

65. On the *Challenger*, see U.S. House of Representatives, 99th Congress, 2nd Session, *The Challenger Accident: Hearings before the Committee on Science and Technology* (Washington, DC: U.S. Government Printing Office, 1986). For the *Columbia*, see the U.S. House of Representatives, 108th Congress, 1st Session, *Columbia Accident Investigation Report: Hearing before the Committee on Science* (Washington, DC: U.S. Government Printing Office, 2004). On both tragedies, see also Julianne G. Mahler, with Maureen Hogan, *Organizational Learning at NASA: The Challenger and Columbia Accidents* (Washington, DC: Georgetown University Press, 2009).

66. For an overview of the literature, see Roger D. Launius, "Interpreting the Moon Landings: Project Apollo and the Historians," *History and Technology* 22, no. 3 (September 2006): 225–255.

67. See the foreword by Neil Armstrong in George Constable and Bob Somerville, *A Century of Innovation: Twenty Engineering Achievements That Transformed Our Lives* (Washington, DC: Joseph Henry Press, 2003), vi–vii.

Chapter 6

1. For the development of the telephone, see David P. Billington and David P. Billington, Jr., *Power Speed and Form: Engineers and the Making of the Twentieth Century* (Princeton, NJ: Princeton University Press, 2006), 35–56. For the development of radio, see ibid., 129–154.

2. Alternating current reverses direction 60 times per second in the United States (50 times per second in Europe) and any flicker is not noticeable. The flicker from blocking current in the negative direction to produce direct current is also not noticeable. For the development of electric power, see Billington and Billington, *Power Speed and Form*, 13–34. For a description of the diode, see J. A. Fleming, "Instrument for Converting Alternating Electric Currents into Continuous Currents," U.S. Patent No. 803,864, November 7, 1905. See also J. A. Fleming, *The Thermionic Valve and Its Developments in Radio-Telegraphy and Telephony*, 2nd ed. (New York: D. Van Nostrand, 1924). Vacuum tubes are called valves in Britain. In the word *diode*, "di" means two and "ode" means pole.

3. For the triode, see L. De Forest, "Space Telegraphy," U.S. Patent No. 879,532, February 18, 1908. See also Lee de Forest, "The Audion: A New Receiver for Wireless Telegraphy," *Transactions of the American Institute of Electrical Engineers* 25 (1906): 735–779; and Robert A. Chipman, "DeForest and the Triode Detector," *Scientific American* 212, no. 3 (March 1965): 92–100. For Lee de Forest (1873–1961), see James A. Hijiya, *Lee de Forest and the Fatherhood of Radio* (Bethlehem, PA: Lehigh University Press, 1992).

4. Bell engineer Harold D. Arnold made the triode practical for use in telephony. See J. E. Brittain, "Harold D. Arnold: A Pioneer in Vacuum-Tube Electronics," *Proceedings of the Institute of Electrical and Electronics Engineers* 86, no. 9 (September 1998): 1895–1896 (hereafter cited as *Proceedings of the IEEE*). Harold S. Black invented negative feedback in 1927, which reduced distortion and made it easier for multiple calls to go over the same line at the same time. On Black, see J. E. Brittain, "Harold S. Black and the Negative Feedback Amplifier," *Proceedings of the IEEE* 85, no. 8 (August 1997): 1335–1336.

5. Armstrong worked independently and his important innovation was the regenerative circuit, which created stronger amplification with triodes. See E. H. Armstrong, "Some Recent Developments in the Audion Receiver," *Proceedings of the Institute of Radio Engineers* 3, no. 4 (September 1915): 215–246. His other early innovation was a new kind of radio called the superheterodyne receiver. See E. H. Armstrong, "The Super-Heterodyne: Its Origin, Development, and Some Recent Improvements," *Proceedings of the Institute of Radio Engineers* 12, no. 5 (October 1924): 539–552. Armstrong invented FM radio in the 1930s. For Edwin Howard Armstrong (1890–1954), see Lawrence Lessing, *Man of High Fidelity: Edwin Howard Armstrong, a Biography* (Philadelphia: J. B. Lippincott, 1956). He was no relation to Neil Armstrong, the astronaut.

6. On the problems of amplification and switching with vacuum tubes, see Lillian Hoddeson, "The Discovery of the Point-Contact Transistor," *Historical Studies in the Physical Sciences* 12, no. 1 (1981): 41–76, especially 45.

7. Science owed its new understanding of semiconductors largely to A. H. Wilson, *Semi-Conductors and Metals: An Introduction to the Electron Theory of Metals* (Cambridge: Cambridge University Press, 1939). On Wilson, see E. H. Sondheimer, "Sir Alan Herries Wilson, 2 July 1906–30 September 1995," in *Biographical Memoirs of Fellows of the Royal Society* (London: The Royal Society, 1999), 547–562. See also Lillian Hartmann Hoddeson, "The Entry of the Quantum Theory of Solids into the Bell Telephone Laboratories: A Case Study of the Industrial Application of Fundamental Science," *Minerva: A Review of Science, Learning, and Policy* 18, no. 3 (autumn 1980): 422–447. Hoddeson's article, on the 1930s, documents how Bell Labs began to give its researchers more freedom to do basic science as part of their work.

8. For early radio-wave detection, see *The Pulse of Radar: The Autobiography of Sir Robert Watson-Watt* (New York: Dial Press, 1959); and Henry E. Guerlac, *Radar in World War II*, 2 vols. (Woodbury, NY: American Institute of Physics, 1987), vol. 1: 3–240. The British RDF system came to be known later as Radio Direction Finding.

9. For the National Defense Research Committee, which became the Office of Scientific Research and Development in 1942, see Irvin Stewart, *Organizing Scientific Research for War: The Administrative History of the Office of Scientific Research and Development* (Boston: Little, Brown, 1948). On Vannevar Bush, see G. Pascal Zachary, *Endless Frontier: Vannevar Bush, Engineer of the American Century* (New York: Free Press, 1997). For the MIT Radiation Laboratory, see Guerlac, *Radar in World War II*, vol. 1: 243–303.

10. On early British RDF and its limitations during the war, see David Zimmerman, *Britain's Shield: Radar and the Defeat of the Luftwaffe* (Phoenix Mill, UK: Sutton Publishing, 2001), 213–224. On the cavity magnetron, see ibid., 227–228. George C. Southworth of Bell Labs found that a crystal rectifier could detect microwaves. Russell L. Ohl, Jack Scaff, and Henry Theurer at Bell Labs discovered the role of impurities in semiconductors. See G. L. Pearson and W. H. Brattain, "History of Semiconductor Research," *Proceedings of the Institute of Radio Engineers* 43, no. 12 (December 1955): 1794–1806; and Michael Riordan and Lillian Hoddeson, *Crystal Fire: The Invention of the Transistor and the Birth of the Information Age* (New York: W. W. Norton, 1997), 88–108.

11. On the refining of silicon at DuPont, see C. Marcus Olson, "The Pure Stuff," *Invention and Technology* 4, no. 1 (spring/summer 1988): 58–63. New methods replaced the DuPont process after the war.

12. For the role of silicon rectifiers in the war, see J. H. Scaff and R. S. Ohl, "The Development of Silicon Crystal Rectifiers for Microwave Radar Receivers," *Bell System Technical Journal* 26, no. 1 (March 1947): 1–30. On the refining of germanium, see Frederick Seitz, "Research on Silicon and Germanium in World War II," *Physics Today* 48, no. 1 (January 1995): 22–27.

13. For Bell Labs as Kelly saw them at this time, see Mervin J. Kelly, "The Bell Telephone Laboratories—An Example of an Institute of Creative Technology," *Proceedings of the Royal Society of*

London, Series A, 203, no. 1074 (October 10, 1950): 287–301. For Kelly's role, see also Riordan and Hoddeson, *Crystal Fire*, 108–110. For a retrospective on Kelly, see *Mervin Joe Kelly 1894–1971: A Biographical Memoir by John R. Pierce* (Washington, DC: National Academy of Sciences, 1975).

14. A chemist, Stanley O. Morgan, also shared responsibility with Shockley for directing the larger solid-state physics group. Morgan assumed most of the administrative duties and supervised people who were not as directly involved in the research that led to the transistor. On Shockley's life and activity up to 1945, see Riordan and Hoddeson, *Crystal Fire*, 21–27, 71–75, 80–87, 104, 107–108. See also John L. Moll, *William Bradford Shockley 1910–1989: A Biographical Memoir* (Washington, DC: National Academy of Sciences, 1995); and Joel N. Shurkin, *Broken Genius: The Rise and Fall of William Shockley, Creator of the Electronic Age* (New York: Macmillan, 2006).

15. On the distinction between n-type and p-type semiconductors, see Michael Riordan and Lillian Hoddeson, "The Origins of the P-N Junction," *IEEE Spectrum* 34, no. 6 (June 1997): 46–51.

16. For Shockley's experiments, see William Shockley, "The Path to the Conception of the Junction Transistor," *IEEE Transactions on Electron Devices*, ED-23, no. 7 (July 1976): 597–620, especially 604–605; and Riordan and Hoddeson, *Crystal Fire*, 110–114.

17. For the life of Brattain, see John Bardeen, *Walter Houser Brattain 1902–1987: A Biographical Memoir* (Washington, DC: National Academy of Sciences, 1994); and Riordan and Hoddeson, *Crystal Fire*, 11–14, 28–29, 55–65, 88–108.

18. For Bardeen's life, see Nick Holonyak, Jr., "John Bardeen 1908–1991," in National Academy of Engineering of the United States of America, *Memorial Tributes* (Washington, DC: National Academy Press, 1993); and David Pines, *John Bardeen 1908–1991: A Biographical Memoir* (Washington, DC: National Academy of Sciences, 2013). For a full biography, see Lillian Hoddeson and Vicki Daitch, *True Genius: The Life and Science of John Bardeen* (Washington, DC: Joseph Henry Press, 2002). On Bardeen's life up to 1945, see also Riordan and Hoddeson, *Crystal Fire*, 15–18, 75–80, 118–119.

19. For these decisions, see W. S. Gorton, "The Genesis of the Transistor," Memorandum for Record, December 27, 1949, reprinted in *A History of Engineering and Science in the Bell System: Physical Sciences (1925–1980)*, ed. S. Millman (Murray Hill, NJ: AT&T Bell Laboratories, 1983), 97–100. Also reprinted in *Proceedings of the IEEE* 86, no. 1 (January 1998): 50–52.

20. For the research of Bardeen and Brattain, see John Bardeen, "Semiconductor Research Leading to the Point-Contact Transistor," Nobel Prize Lecture, December 11, 1956, in *Nobel Lectures in Physics* (Singapore: World Scientific Publishing, 1998), 3: 318–341. The Nobel Lecture by Walter Brattain agreed with Bardeen and provided additional detail in ibid., 377–384.

21. See again the two Nobel Lectures cited in note 20 above, and Riordan and Hoddeson, *Crystal Fire*, 128–136.

22. In addition to the Nobel Lectures of Bardeen and Brattain, see again Gorton, "Genesis of the Transistor"; and J. Bardeen and W. H. Brattain, "The Transistor, A Semiconductor Triode," *Physical*

Review 74, no. 2 (July 15, 1948): 230–231; reprinted in *Proceedings of the IEEE* 86, no. 1 (January 1998): 29–30. See also Riordan and Hoddeson, *Crystal Fire*, 115–141; and Michael Riordan and Lillian Hoddeson, "Minority Carriers and the First Two Transistors," in *Facets: New Perspectives on the History of Semiconductors*, ed. Andrew Goldstein and William Aspray (New Brunswick, NJ: IEEE Center for the History of Electrical Engineering, 1997), 1–33. For the term "transistor," see John R. Pierce, "The Naming of the Transistor," *Proceedings of the IEEE* 86, no. 1 (January 1998): 37–45.

23. Professor Karl Lark-Horovitz led the Purdue research and Ralph Bray was the graduate student. For the work at Purdue, see Ralph Bray, "The Invention of the Point-Contact Transistor: A Case Study in Serendipity," in *Silicon Materials Science and Technology: Proceedings of the Eighth International Symposium on Silicon Materials Science and Technology*, 2 vols. (Pennington, NJ: The Electrochemical Society, 1998), 1: 143–156; and "Interview of Ralph Bray by Paul Henriksen, May 14, 1982," in the Niels Bohr Library & Archives, American Institute of Physics, College Park, MD. "I simply didn't think so much in terms of devices," observed Bray in the interview. "There was never an emphasis in the academic world in terms of making devices. . . . And it wouldn't have occurred to me that a triode might even be a worthwhile endeavor." For the Purdue research on germanium and its importance to the decision of the Bell scientists to switch from silicon to germanium in the final stage of their transistor work, see Paul W. Henriksen, "Solid State Physics Research at Purdue," *Osiris*, 2nd series, no. 3 (1987): 237–260.

24. On the difficulty with point-contact transistors, see Riordan and Hoddeson, *Crystal Fire*, 168–170. For Shockley's reaction to his exclusion from the patent claim for the point-contact transistor, see ibid., 144–146. Julius Lilienfeld had an earlier patent for using a field effect to control a current in a semiconductor, filed October 8, 1926, and granted January 28, 1930, as U.S. Patent No. 1,745,175. It is not clear whether he tested the idea; if he had, he might have encountered the same problem that Shockley found in his 1945 experiments.

25. For the development of his transistor, see William Shockley, "The Path to the Conception of the Junction Transistor," *IEEE Transactions on Electron Devices*, ED-23, no. 7 (July 1976): 597–620; and Riordan and Hoddeson, *Crystal Fire*, 142–144, 148–167, who also describe the January 1948 experiment confirming Shockley's idea. The experiment was planned before Shockley disclosed his idea. Instead of placing two contacts on the same surface, a researcher placed the second contact on the opposite side of a germanium piece and amplified a current, showing that amplification was not just a surface effect. See also W. Shockley, "The Theory of P-N Junctions in Semiconductors and P-N Junction Transistors," *Bell System Technical Journal* 28 (1949): 435–489. Shockley went on to publish a definitive textbook on semiconductor electronics, *Electrons and Holes in Semiconductors* (New York: Van Nostrand, 1950).

26. Purified silicon and germanium were polycrystalline, i.e., the atoms formed a regular pattern inside regions but the regions did not align with each other exactly. Gordon Teal at Bell Labs found a way to produce monocrystalline germanium, in which the atoms throughout the material lined up regularly, eliminating distinct regions. See G. K. Teal, M. Sparks, and E. Buehler, "Growth of Germanium Single Crystals Containing P-N Junctions," *Physical Review* 81, no. 4 (February 1951): 637 (one page); and Gordon K. Teal, "Single Crystals of Germanium and

Silicon—Basic to the Transistor and Integrated Circuit," *IEEE Transactions on Electron Devices* 23, no. 7 (July 1976): 621–639. See also Riordan and Hoddeson, *Crystal Fire*, 172–175, 178–180.

27. For later improvements in the design and manufacturing of transistors, see Ian Ross, "The Invention of the Transistor," *Proceedings of the IEEE* 86, no. 1 (January 1998): 7–28, especially 15–22. For the role of RCA in making transistors practical to manufacture, see also Hyungsub Choi, "The Boundaries of Industrial Research: Making Transistors at RCA, 1948–1960," *Technology and Culture* 48, no. 4 (October 2007): 758–782.

28. On the early market for transistors, see C. Mark Melliar-Smith, M. G. Borrus, D. E. Haggan, T. Lowrey, A. S. G. Vincentelli, and W. W. Troutman, "The Transistor: An Invention Becomes Big Business," *Proceedings of the IEEE* 86, no. 1 (January 1998): 86–110; and Riordan and Hoddeson, *Crystal Fire*, 195–224. For the Regency radio, see Paul D. Davis, "The Breakthrough Breadboard Feasibility Model: The Development of the First All-Transistor Radio," *Southwestern Historical Quarterly* 97, no. 1 (July 1993): 56–80, introduction by Diana Kleiner. For a more complete overview of portable radios, both before and after the transistor, see Michael Brian Schiffer, *The Portable Radio in American Life* (Tucson: University of Arizona Press, 1991), 161–201. For the Japanese entry into the market for transistor radios, see ibid., 202–223.

29. For the MOSFET transistor, see D. Kahng and M. M. Atalla, "Silicon-Silicon Dioxide Field Induced Surface Devices," in *IRE-AIEE Solid State Device Research Conference* (Pittsburgh, PA: Carnegie Institute of Technology, 1960); and D. Kahng, "A Historical Perspective on the Development of MOS Transistors and Related Devices," *IEEE Transactions on Electron Devices* 23, no. 7 (July 1976): 655–657. The metal-oxide referred to a coating left in place during the manufacturing process to protect the substrate material (see chapter 7). For the development of the MOS device, see also Ross Knox Bassett, *To the Digital Age: Research Labs, Start-up Companies, and the Rise of MOS Technology* (Baltimore, MD: Johns Hopkins University Press, 2002).

30. On Shockley's later life, see Shurkin, *Broken Genius*, 164–189, 190–270.

31. On Bardeen's later life, see Hoddeson and Daitch, *True Genius*, 142 passim.

32. For his 1945 report, see Vannevar Bush, *Science, the Endless Frontier: A Report to the President* (Washington, DC: U.S. Government Printing Office, 1945). Quotes are from ibid., 18–19.

33. For the development of penicillin, see Gwyn MacFarlane, *Howard Florey: The Making of a Great Scientist* (New York: Oxford University Press, 1979).

34. The engineering mission is clear from the title of Bardeen and Brattain, "The Transistor, A Semiconductor Triode." For the importance of the wartime engineering work on semiconductors, see M. Gibbons and C. Johnson, "Science, Technology, and the Development of the Transistor," in *Science in Context: Readings in the Sociology of Science*, ed. Barry Barnes and David Edge (Cambridge, MA: MIT Press, 1982), 177–185. For critical perspectives on the Vannevar Bush thesis from differing points of view, see *Science for the Twenty-First Century: The Bush Report Revisited*, ed. Claude Barfield (Washington, DC: AEI Press, 1997); and Donald E. Stokes, *Pasteur's Quadrant: Basic Science and Technological Innovation* (Washington, DC: Brookings Institution Press, 1997).

35 For the later history of Bell Labs, see Jon Gertner, *The Idea Factory: Bell Labs and the Great Age of American Innovation* (New York: Penguin Press, 2012).

Chapter 7

1. For an account of Kilby and Noyce, see T. R. Reid, *The Chip: How Two Americans Invented the Microchip and Launched a Revolution* (New York: Random House, 2001). The early life of Jack Kilby is documented in Folders 1–10 in Box 1 of the Jack Kilby Papers, DeGolyer Library, Southern Methodist University, Dallas, TX. Cited hereafter as the Kilby Papers (SMU). For an account of his early years by a colleague and close friend, see Ed Millis, *Jack St. Clair Kilby: A Man of Few Words* (Dallas, TX: Ed Millis Books, 2008), 1–34. Kilby's boyhood amateur radio call sign was W9GTY.

2. Millis, *Jack St. Clair Kilby*, 35–42. On Centralab and Kilby's work there, see also Reid, *The Chip*, 69–72. For the *Annual Reports* of the Globe-Union Company from 1954–1961, see Folder 24, Box 1, Kilby Papers (SMU). For his reasons to leave Centralab, see Jack S. Kilby to Robert L. Wolff, carbon copy [1958], in Folder 5, Box 80, Jack Kilby Papers, Manuscript Division, U.S. Library of Congress, Washington, DC. Hereafter cited as the Kilby Papers (LC). See also J. S. Kilby to Harry C. Laur, March 10, 1958, in the same folder. Laur was personnel administrator at Texas Instruments (TI), and in his letter to Laur, Kilby gave a more detailed explanation as part of his application for a position. TI hired him in a letter from Willis Adcock to Jack S. Kilby, March 31, 1958, in the same folder. Adcock managed the group that Kilby joined. The same folder contains a letter of rejection from IBM dated April 16, 1958; and a letter from Motorola dated April 22, expressing regret that Kilby had declined their offer.

3. For the early years of Robert Noyce, including his time at Grinnell College, see Leslie Berlin, *The Man behind the Microchip: Robert Noyce and the Invention of Silicon Valley* (New York: Oxford University Press, 2005), 7–27. The papers of Robert Noyce are in Collection M1490, Department of Special Collections and University Archives, Stanford University Library, Palo Alto, California.

4. On the influence of Grant Gale, see Berlin, *The Man behind the Microchip*, 24–27. For Noyce at MIT and Philco, see ibid., 28–52.

5. For Shockley's telephone call to Noyce, see Reid, *The Chip*, 87. Reid interviewed Kilby and Noyce for his book.

6. For Shockley's departure from Bell Labs and company in Mountain View, California, see Joel N. Shurkin, *Broken Genius: The Rise and Fall of William Shockley, Creator of the Electronic Age* (New York: Macmillan, 2006), 163–189. For Shockley's recruitment of Noyce and the other younger men, their difficulties working under Shockley, the resignation of Noyce and his colleagues, and the formation of Fairchild Semiconductor, see Berlin, *The Man behind the Microchip*, 53–81. On Sherman Fairchild, see Steven W. Usselmann, "Sherman Mills Fairchild," in *Dictionary of American Biography*, Supplement 9 (New York: Scribner, 1975), 273–274.

7. See J. A. Morton and W. J. Pietenpol, "The Technological Impact of Transistors," *Proceedings of the Institute of Radio Engineers* 46, no. 6 (June 1958): 955–959. Morton and Pietenpol noted that

the "tyranny of numbers" was a threat to all electron devices, but in their paper they discussed the problem as one affecting vacuum tubes, not transistors.

8. For a survey of the main approaches to making electronics smaller, see E. F. Horsey and P. J. Franklin, "Status of Microminiaturization," *IRE Transactions on Component Parts* 9, no. 1 (March 1962): 3–19. For the interest of the U.S. armed forces in miniaturization, and the various research programs that attempted to achieve it in the 1950s, see Christophe Lécuyer and David C. Brock, *Makers of the Microchip: A Documentary History of Fairchild Semiconductor* (Cambridge, MA: MIT Press, 2010), 34–38.

9. On Project Tinkertoy, see R. L. Henry and C. C. Rayburn, "Mechanized Production of Electronic Equipment," *Electronics* 26, no. 12 (December 1953): 160–165. See also the summary report in ibid., 5–6, of the same issue. The name "Tinkertoy" came from a popular children's toy with which objects could be made by assembling wooden sticks.

10. On the Army Micro-Module, see S. F. Danko, W. L. Doxey, and J. P. McNaul, "The Micro-Module: A Logical Approach to Microminiaturization," *Proceedings of the Institute of Radio Engineers* 47, no. 5 (May 1959): 894–904; and William R. Stevenson, *Miniaturization and Microminiaturization of Army Communications—Electronics, 1946–1964* (Fort Monmouth, NJ: U.S. Army Electronics Command, 1966), 149–174.

11. See A. von Hippel, "Molecular Engineering," *Science* 123, no. 3191 (February 24, 1956): 315–317.

12. The "molecular electronics" envisioned at this time was an idea of materials engineering and differed from later "molecular electronics" ("moletronics") or circuits at molecular scales. For an overview of both, see Hyungsub Choi and Cyrus C. M. Mody, "The Long History of Molecular Electronics," *Social Studies of Science* 39, no. 1 (February 2009): 11–50. For a description of the Westinghouse program, see S. W. Herwald, "The Concepts and Capabilities of Molecular Electronics," *Westinghouse Engineer* 20, no. 3 (May 1960): 66–70. A shorter description is included in E. F. Horsey and P. J. Franklin, "Status of Microminiaturization," *IRE Transactions on Component Parts* 9, no. 1 (March 1962): 17. For concern about the slow pace of the Westinghouse work, see R. D. Alberts, "Microminiaturization and Molecular Electronics," in *Electronics Reliability & Microminiaturization* (Oxford: Pergamon Press, 1962), 1: 233–238, especially 237. Colonel Alberts headed the Wright Air Development Center at Wright-Patterson Air Force Base, in which capacity he supervised electronic miniaturization research for the U.S. Air Force.

13. See Jack S. Kilby, "The Invention of the Integrated Circuit," *IEEE Transactions on Electron Devices*, ED-23, no. 7 (July 1976): 648–654. For his mention of Dummer, see ibid., 648–649. See also G. W. A. Dummer, "Electronic Components in Great Britain," in *Proceedings of the Symposium on Progress in Quality Electronic Components*, Washington, DC, May 6, 1952, 15–20. Kilby did not know of Dummer's work until later. See p. 10 in the transcript of Kilby's interview with Buzz Selby, a colleague at Texas Instruments, September 14, 1976, in Folder 6, Box 95, Kilby Papers (LC).

14. For Haggerty, see J. Erik Jonsson, "Patrick Eugene Haggerty," in National Academy of Engineering of the United States of America, *Memorial Tributes* (Washington, DC: National Academy

Press, 1984), 2: 101–105. For Teal, see Don W. Shaw, "Gordon K. Teal 1907–2003," in ibid., 12: 310–313. The publisher is listed in vol. 12 as "National Academies Press." See also again Teal, "Single Crystals of Germanium and Silicon—Basic to the Transistor and Integrated Circuit," *IEEE Transactions on Electron Devices*, ED-23, no. 7 (July 1976): 621–639.

15. For an account of Kilby's early work at Texas Instruments, see Jack S. Kilby, "Turning Potential into Realities: The Invention of the Integrated Circuit," Nobel Lecture, December 8, 2000, in *Nobel Lectures: Physics*, ed. Gösta Ekspong (Singapore: World Scientific, 2002), 471–485. Quote is from ibid., 479. See also Reid, *The Chip*, 73–76.

16. For Kilby's concern with cost, see pp. 7–8 in the transcript of his interview with Buzz Selby, September 14, 1976, in Folder 6, Box 95, Kilby Papers (LC). "Any engineer at Motorola, Zenith, etc. who worked on a [television] set design could tell you to a penny what it would cost to produce that thing," noted Kilby, "and . . . in the consumer end of the business those practices were bred into you."

17. For his breakthrough insight, see Kilby, "Turning Potential into Realities"; and Reid, *The Chip*, 76–78. Quote is from Jack Kilby, entry for July 24, 1958, Laboratory Notebook, p. 8. Texas Instruments placed the notebook online, where it was retrieved by this author on October 3, 2010. The document is no longer in the Kilby section of the TI website and does not appear in the finding aids to the Kilby Papers in the DeGolyer Library at SMU or to the Kilby papers in the U.S. Library of Congress. It is presumably still held by Texas Instruments.

18. For Kilby's prototypes, see Kilby, "The Invention of the Integrated Circuit," *IEEE Transactions on Electron Devices*, ED-23, no. 7 (July 1976): 648–654. See also Reid, *The Chip*, 78–80.

19. For Kilby's original salary, see Willis Adcock to Jack S. Kilby, March 31, 1958, Folder 5, Box 80, Kilby Papers (LC). For his raise in salary, in the same folder, see Mark Sheperd, Jr., to J. S. Kilby, December 1, 1958.

20. For his patent, see J. S. Kilby, "Miniaturized Electronic Circuits," U.S. Patent No. 3,138,743, filed February 6, 1959, issued June 23, 1964. Concerned that other firms might be coming close to the idea, Texas Instruments decided that speed in filing was more important than trying to work out more fully how to manufacture the device without wires. See Reid, *The Chip*, 104–106.

21. For the launching of Fairchild and the successful delivery of the 100 transistors to IBM, see Berlin, *The Man behind the Microchip*, 82–96. For the seminal early years of Fairchild Semiconductor, see also Christophe Lécuyer, *Making Silicon Valley: Innovation and the Growth of High Tech, 1930–1970* (Cambridge, MA: MIT Press, 2006), 129–167; and Lécuyer and Brock, *Makers of the Microchip*, 9–44. The transistor fabrication method was called double diffusion.

22. For the process of silicon dioxide masking, see L. Derick et al., "Oxidation of Semiconductive Surfaces for Controlled Diffusion," U.S. Patent No. 2,802,760, filed December 2, 1955, issued August 13, 1957. The inventors were Lincoln Derick and Carl J. Frosch. For the invention of the process at Bell Labs, see Michael Riordan and Lillian Hoddeson, *Crystal Fire: The Invention of the Transistor and the Birth of the Information Age* (New York: W. W. Norton, 1997), 217–223.

23. For the planar process, see J. A. Hoerni, "Method of Manufacturing Semiconductor Devices," U.S. Patent No. 3,025,589, filed May 1, 1959, issued March 20, 1962; and J. Hoerni, "Semiconductor Device," U.S. Patent No. 3,064,167, filed May 1, 1959, issued November 13, 1962. See also Michael Riordan, "The Silicon Dioxide Solution," *IEEE Spectrum* 44, no. 12 (December 2007): 51–56.

24. For a reproduction of Noyce's laboratory notebook entry of January 23, 1959, see Lecuyer and Brock, *Makers of the Microchip*, 151–155. The original notebook pages are numbered 70–74.

25. For his account of integrated circuit development, see Robert N. Noyce, "Microelectronics," *Scientific American* 237, no. 3 (September 1977): 63–69. For his patent, see R. N. Noyce, "Semiconductor Device-and-Lead Structure," U.S. Patent No. 2,981,877, filed July 30, 1959, issued April 25, 1961. See also Berlin, *The Man behind the Microchip*, 97–111; and Lécuyer and Brock, *Makers of the Microchip*, 34–39, 150–161.

26. Quote is from Reid, *The Chip*, 65–66. For Noyce's approach, see Berlin, *The Man behind the Microchip*, 97–98.

27. On the litigation between Texas Instruments and Fairchild Semiconductor, and their agreement to cross-license their patents in 1966, see Reid, *The Chip*, 96–117. For the decision in the U.S. Court of Customs and Patent Appeals, see the *Federal Reporter*, 2nd Series (St. Paul, MN: West Publishing, 1924–1993), vol. 416, 1391. The appeals court credited Fairchild with having originated a process for making the chip. The U.S. Supreme Court declined to hear the case on October 12, 1970. Noyce used a method of isolating the different parts of the circuit conceived in 1958 by Kurt Lehovec, an engineer at Sprague Electric near Boston, who used junctions of p-type and n-type material to keep electrical currents in one place from interfering with them in another (Kilby had accomplished this isolation in his prototypes by inscribing trenches or by spacing apart the component parts). Later integrated circuits used doping in different places to create junctions. For Lehovec's work, see K. Lehovec, "Multiple Semiconductor Assembly," U.S. Patent No. 2,029,366, filed April 22, 1959, issued April 10, 1962. Noyce did not know of Lehovec's work at the time of his own insight in January 1959. See Lécuyer and Brock, *Makers of the Microchip*, 39. Once he learned of the Lehovec patent, Noyce revised his patent to differentiate it.

28. Quote is from Kilby, "Turning Potential into Realities," 482.

29. See Bernard T. Murphy, Douglas E. Haggan, and William Troutman, "From Circuit Miniaturization to the Scalable IC," *Proceedings of the IEEE* 88, no. 5 (May 2000): 690–703.

30. The contract for the computer on the Minuteman II missile went to Autonetics, the aviation electronics ("avionics") division of North American Aviation, which gave Texas Instruments an order for integrated circuits. For the early purchases of integrated circuits by the armed forces and the space program, see Barry Miller, "Microcircuitry Production Growth Outpaces Applications," *Aviation Week and Space Technology* 81, no. 20 (November 16, 1964): 76–87. See also R. C. Platzek and J. S. Kilby, "MINUTEMAN Integrated Circuits—A Study in Combined Operations," *Proceedings of the IEEE* 52, no. 12 (December 1964): 1669–1678.

31. On the Apollo computer, see again Eldon C. Hall, *Journey to the Moon: A History of the Apollo Guidance Computer* (Reston, VA: Institute of Aeronautics and Astronautics, 1996); and David Mindell, *Digital Apollo: Human and Machine in Six Lunar Landings* (Cambridge, MA: MIT Press, 2008). Fairchild Semiconductor withdrew from supplying the space program in the mid-1960s, after having established the integrated circuit as a viable innovation. The requirements of the Apollo program demanded a narrowly defined product, and both TI and Fairchild were anxious to serve a more diversified market. Raytheon and Philco (now part of the Ford Motor Company) took over supplying the needed microchips to NASA in the late 1960s and early 1970s using the Fairchild design. Each Apollo mission carried several thousand integrated circuits in each of two computers, one in the command ship and one in the lunar lander. See Hall, *Journey to the Moon*, 34. For the role of NASA in launching the integrated circuit, see also Andrew Butrica, "NASA's Role in the Manufacture of Integrated Circuits," in *Historical Studies in the Societal Impact of Space Flight*, ed. Steven J. Dick (Washington, DC: NASA History Program Office, 2015), 149–249. Butrica also describes the role of military purchases for the Minuteman missile. The Apollo computers used magnetic memory, an older technology, for memory storage.

32. For the handheld calculator, see Kilby et al., "Miniature Electronic Calculator," U.S. Patent No. 3,819,921, filed December 21, 1972, issued June 25, 1974. The patent was a revised version of one originally filed in 1967. See also Kathy B. Hamrick, "The History of the Hand-Held Electronic Calculator," *The American Mathematical Monthly* 103, no. 8 (October 1996): 633–639.

33. For Moore's initial skepticism, see Lécuyer, *Making Silicon Valley*, 214. For his article, see Gordon E. Moore, "Cramming More Components onto Integrated Circuits," *Electronics* 38, no. 8 (April 19, 1965): 114–117. Reprinted in *Proceedings of the IEEE* 86, no. 1 (January 1998): 82–85. See also E. Mollick, "Establishing Moore's Law," *IEEE Annals of the History of Computing* 28, no. 3 (July–September 2006): 62–75.

34. For the origins of Silicon Valley, see again Lécuyer, *Making Silicon Valley*.

35. For Frederick Emmons Terman and Stanford University, see C. Stewart Gillmor, *Fred Terman at Stanford: Building a Discipline, a University, and Silicon Valley* (Stanford, CA: Stanford University Press, 2004).

36. See F. Terman, "Engineering Growth and the Community," in *The World of Engineering*, ed. John R. Whinnery (New York: McGraw-Hill, 1965), 285–297. Quote is from ibid., 286–287.

37. Terman, "Engineering Growth and the Community," 293.

38. For Noyce's later career in business and his mentorship of younger entrepreneurs, see Berlin, *The Man behind the Microchip*, 207–256. For his warning against American complacency, see R. N. Noyce, "False Hopes and High-Tech Fiction," *Harvard Business Review* 68, no. 1 (January–February 1990): 31–32, 36. For the foreign challenge to the domestic American consumer electronics industry in the 1970s and 1980s, see Alfred D. Chandler, *Inventing the Electronic Century: The Epic Story of the American Consumer Electronic and Computer Industries* (Cambridge, MA: Harvard University Press, 2005), written with the assistance of Takashi Hikino and Andrew von Nordenflycht.

39. For Kilby's later life, see Millis, *Jack St. Clair Kilby*, 65–98; and Thomas Haigh, "Jack Kilby (1923–2005)," *IEEE Annals of the History of Computing* 29, no. 1 (January–March 2007): 90–95.

40. For Noyce's leadership in organizing the semiconductor industry, see Robert N. Noyce, "Competition and Cooperation—A Prescription for the Eighties," *Research Management* 25, no. 2 (March 1982): 13–17; and Berlin, *The Man behind the Microchip*, 257–304.

41. For Kilby's tribute to Noyce, see Reid, *The Chip*, 264. For the Townes quote, see Kilby, "Turning Potential into Realities," 474.

Chapter 8

1. On the flip-flop circuit, see W. H. Eccles and F. W. Jordan, "A Trigger Relay," *Radio Review* 1, no. 10 (October 1919): 143–146.

2. For the British machine, see Thomas H. Flowers, "The Design of Colossus," *Annals of the History of Computing* 5, no. 3 (July–September 1983): 239–252. On Alan Mathison Turing (1912–1954), see Andrew Hodges, *Alan Turing: The Enigma* (Princeton, NJ: Princeton University Press, 1983; 2014). Tragically, Turing committed suicide in 1954 after a conviction for private homosexual activity, which was outlawed at the time.

3. For the design and working of the ENIAC machine, see J. P. Eckert et al., U.S. Patent No. 3,120,606, filed June 26, 1947, granted February 4, 1964. For a historical account of the ENIAC and its early successors, see Nancy Stern, *From ENIAC to UNIVAC: An Appraisal of the Eckert-Mauchly Computers* (Bedford, MA: Digital Press, 1981). See also W. Barkley Fritz, "ENIAC—A Problem Solver," *IEEE Annals of the History of Computing* 16, no. 1 (January 1994): 25–45. For the contribution of women to the programming, see W. Barkley Fritz, "The Women of ENIAC," *IEEE Annals of the History of Computing* 18, no. 3 (January 1996): 13–28. The men on the team considered programming to be like secretarial work, which employed mostly women. It turned out to be just as essential as hardware design. The question of who invented the computer was later disputed. Mauchly and Eckert borrowed some ideas from an Iowa engineer, John Atanosoff, but the Philadelphia engineers were the first to design and operate a fully general-purpose electronic computer. For the dispute, see Charles E. McTiernan, "The ENIAC Patent," *IEEE Annals of the History of Computing* 20, no. 2 (February 1998): 54–58, 80. See also Stern, *From ENIAC to UNIVAC*, 33–34.

4. For a technical introduction to computing theory, see Cullen Schaffer, *Principles of Computer Science* (Englewood Cliffs, NJ: Prentice-Hall, 1988). For historical overviews of the modern electronic computer, see Paul E. Ceruzzi, *A History of Modern Computing*, 2nd ed. (Cambridge, MA: MIT Press, 2003); and Martin Campbell-Kelly and William Aspray, *Computer: A History of the Information Machine*, 2nd ed. (Boulder, CO: Westview Press, 2004). For the idea of a stored program and the division of a computer into five parts, see John von Neumann, *First Draft of a Report on the EDVAC, June 30, 1945* (Philadelphia: Moore School of Engineering, University of Pennsylvania, 1945), reprinted in *IEEE Annals of the History of Computing* 15, no. 4 (April 1993): 27–75. EDVAC stood for Electronic Discrete Variable Automatic Computer and was a successor to ENIAC. See also M. D. Godfrey and D. F. Hendry, "The Computer as Von Neumann Planned It,"

IEEE Annals of the History of Computing 15, no. 1 (January 1993): 11–21; and Michael R. Williams, "The Origins, Uses and Fate of the EDVAC," in the same issue, 22–38. Who originated the idea of a stored program has been disputed. See Stern, *From ENIAC to UNIVAC*, 74–75. It appears to have emerged in group discussions among members of the ENIAC group during the war, including von Neumann, Mauchly, and Eckert.

5. See Claude E. Shannon, "A Symbolic Analysis of Relay and Switching Circuits," *Transactions of the American Institute of Electrical Engineers* 57, no. 12 (December 1938): 713–723. For his later paper, see C. E. Shannon, "A Mathematical Theory of Communication," *The Bell System Technical Journal* 27, no. 3 (July 1948): 379–423, and no. 4 (October 1948): 623–656. For Shannon's life, see Jimmy Soni and Rob Goodman, *A Mind at Play: How Claude Shannon Invented the Information Age* (New York: Simon & Schuster, 2017). Shannon credited a Princeton professor, John Tukey, with inventing the term "binary digit," or "bit."

6. UNIVAC stood for Universal Automatic Computer. On the later work of Mauchly and Eckert, see Stern, *From ENIAC to UNIVAC*, 87–159.

7. On IBM and its early computers, see Emerson W. Pugh, *Building IBM: Shaping an Industry and Its Technology* (Cambridge, MA: MIT Press, 1995); and Charles J. Bashe, Lyle R. Johnson, John H. Palmer, and Emerson W. Pugh, *IBM's Early Computers* (Cambridge, MA: MIT Press, 1986).

8. On the IBM System/360, see Emerson W. Pugh, Lyle R. Johnson, and John H. Palmer, *IBM's 360 and Early 370 Systems* (Cambridge, MA: MIT Press, 1991). For its size, see the IBM System/360 Installation Manual: Physical Planning (Armonk, NY: IBM Systems Reference Library, 1974), E-01. For the rise of mainframe computers more generally, see Ceruzzi, *A History of Modern Computing*, 13–78. For the coming of smaller minicomputers, see ibid., 109–141.

9. On Hopper's work, see Grace Murray Hopper, "The Education of a Computer," *Proceedings of the Association for Computing Machinery, May 2 and 3, 1952* (Pittsburgh: Association for Computing Machinery, 1952), 243–249; and Grace M. Hopper, "Compiling Routines," *Computers and Automation* 2, no. 4 (May 1953): 1–5. For Hopper, who retired as a rear admiral in the U.S. Naval Reserve, see Kathleen Broome Williams, *Grace Hopper: Admiral of the Cyber Sea* (Annapolis: Naval Institute Press, 2004); and Kurt W. Beyer, *Grace Hopper and the Invention of the Information Age* (Cambridge, MA: MIT Press, 2009).

10. On early computer software, see Ceruzzi, *A History of Modern Computing*, 79–108; and Martin Campbell-Kelly, *From Airline Reservations to Sonic the Hedgehog: A History of the Software Industry* (Cambridge, MA: MIT Press, 2003), 29–55. COBOL is short for Common Business-Oriented Language. On FORTRAN (Formula Translator), see J. W. Backus and W. P. Heising, "Fortran," *IEEE Transactions on Electronic Computers* EC-13, no. 4 (August 1964): 382–385; and on BASIC (Beginner's All-purpose Symbolic Instruction Code), see John G. Kemeny and Thomas E. Kurtz, *BASIC Programming* (New York: John Wiley and Sons, 1967). See also Campbell-Kelly and Aspray, *Computer*, 34–36, 187–189.

11. On SAGE (Semi-Automatic Ground Environment), see Kent C. Redmond and Thomas M. Smith, *From Whirlwind to MITRE: The R&D Story of the SAGE Air Defense Computer* (Cambridge,

MA: MIT Press, 2000); and on SABRE (Semi-Automatic Business Research Environment), see R. W. Parker, "The SABRE System," *Datamation* 11, no. 9 (September 1965): 49–52; and Campbell-Kelly, *From Airline Reservations to Sonic the Hedgehog*, 41–45.

12. On Philo Taylor Farnsworth (1906–1971), see Donald G. Godfrey, *Philo T. Farnsworth: The Father of Television* (Salt Lake City: University of Utah Press, 2001). For his television design, see P. T. Farnsworth, "Television System," U.S. Patent No. 1,773,980, filed January 7, 1927, and granted August 26, 1930.

13. On Vladimir Kosma Zworykin (1889–1982), see Albert Abramson, *Zworykin: Pioneer of Television* (Urbana: University of Illinois Press, 1995). For the Zworykin system, see V. K. Zworykin, "Television System," U.S. Patent Nos. 2,022,450 and 2,141,059, originally filed as one patent, December 29, 1923, and granted as two on November 26, 1935, and December 20, 1938.

14. For Sarnoff's 1939 demonstration of television, see Eugene Lyons, *David Sarnoff: A Biography* (New York: Harper & Row, 1966), 216; and for RCA's interest in television, see ibid., 204–220. On the growth of television to 1941, see Joseph H. Udelson, *The Great Television Race: A History of the American Television Industry, 1925–1941* (Tuscaloosa: University of Alabama Press, 1982). For the growth of television networks and TV viewing in post-1945 America, see Douglas Gomery, "Television," in *A Companion to American Technology*, ed. Carroll Pursell (Malden, MA: Blackwell, 2005), 321–339.

15. For the engineering of television, see V. K. Zworykin and G. A. Morton, *Television: The Electronics of Image Transmission in Color and Monochrome*, 2nd ed. (New York: John Wiley and Sons, 1954). For the use of transistors, see Gerald L. Hansen, *Introduction to Solid-State Television Systems: Color and Black & White* (Englewood Cliffs, NJ: Prentice-Hall, 1969).

16. On the spread of cable and satellite television, see Patrick R. Parsons and Robert M. Frieden, *The Cable and Satellite Television Industries* (New York: Allyn and Bacon, 1998).

17. See Kent C. Redmond and Thomas M. Smith, *Project Whirlwind: The History of a Pioneer Computer* (Bedford, MA: Digital Press, 1980), 216. Computers such as the PDP-1 began to use visual displays in the 1960s.

18. For the founding and early focus of Intel, see Christophe Lécuyer, *Making Silicon Valley: Innovation and the Growth of High Tech, 1930–1970* (Cambridge, MA: MIT Press, 2006), 279–287. For the new crisis in hardware design, see Robert N. Noyce and Marcian E. Hoff, Jr., "A History of Microprocessor Development at Intel," *IEEE Micro* 1, no. 1 (February 1981): 8–21.

19. For the Intel microprocessor, see Marcian Edward Hoff, Jr., Stanley Mazor, and Federico Faggin, "Memory System for a Multi-Chip Digital Computer," U.S. Patent No. 3,821,715, filed January 22, 1973, granted June 28, 1974. See also Robert N. Noyce and Marcian E. Hoff, Jr., "A History of Microprocessor Development at Intel," *IEEE Micro* 1, no. 1 (February 1981): 8–21; and Federico Faggin, Marcian E. Hoff, Jr., Stanley Mazor, and Masatoshi Shima, "The History of the 4004," *IEEE Micro* 16, no. 6 (December 1996): 10–20. See also Ceruzzi, *A History of Modern Computing*, 217–221. The 4004 was the first commercial microprocessor. Intel sold it with three other chips to provide the functions of a small computer.

20. For Intel's initial reluctance, and then decision to support and develop the microprocessor, see again Noyce and Hoff, "A History of Microprocessor Development at Intel.," *IEEE Micro*, 1, No. 1 (February 1981): 8–21. William Aspray, in "The Intel 4004 Microprocessor: What Constituted Invention?," *IEEE Annals of the History of Computing*, 19, no. 3 (March 1997): 4–15, also credits the prior work of Japanese calculator engineers that helped to stimulate the design of the Intel 4004.

21. See Carver Mead and Lynn Conway, *Introduction to VLSI Systems* (Reading, MA: Addison-Wesley, 1979). For her career, see Lynn Conway, "Reminiscences of the VLSI Revolution: How a Series of Failures Triggered a Paradigm Shift in Digital Design," *IEEE Solid State Circuits*, 4, no. 4 (fall 2012): 8–31.

22. MITS was short for Micro Instrumentation and Telemetry Systems, which began as a producer of telemetry for model aircraft. For the Altair 8800, see H. Edward Roberts and William Yates, "Exclusive! Altair 8800: The Most Powerful Minicomputer Project Ever Presented—Can Be Built for Under under $400," *Popular Electronics*, 7, no. 1 (January 1975): 33–38; and "Build the Altair 8800 Minicomputer Part Two," *Popular Electronics* 7, no. 2 (February 1975): 56–58. On the Altair and MITS, see Campbell-Kelly and Aspray, *Computer*, 213–214; and Ceruzzi, *A History of Modern Computing*, 2nd ed., 226–232.

23. For the early life of Steve Wozniak, see Steve Wozniak with Gina Smith, *iWoz: Computer Geek to Cult Icon* (New York: W. W. Norton, 2006), 1–92.

24. For Steve Jobs, see Daniel Morrow, "Interview with Steve Jobs," Computer History Collection, National Museum of American History, Smithsonian Institution, Washington, DC, 1995. For his early life and relationship with Steve Wozniak, see also Wozniak, *iWoz*, pp. 93–149; and Walter Isaacson, *Steve Jobs* (New York: Simon & Schuster, 2011), 1–63.

25. On the Apple I computer, see the *Apple-I Operation Manual* (Palo Alto, CA: Apple Computer Company, [1976]). See also "Apple Introduces the First Low Cost Microcomputer System with a Video Terminal . . . ," 2 pp., in Apple Computer, Inc., Records, 1977–1998, M1007, Series 3, Box 12, Folder 14, Stanford University Library, Palo Alto, California. For the Apple I and the formation of the Apple Computer Company, see Wozniak, *iWoz*, 150–185; and Isaacson, *Steve Jobs*, 63–70. Wozniak designed the Apple I to use a form of the programming language BASIC.

26. On the Apple II, see Stephen G. Wozniak, "Microcomputer for Use with Video Display," U.S. Patent No. 4,136,359, filed April 11, 1977, granted January 23, 1979; and Stephen Wozniak, "The Apple-II: System Description," *Byte: The Small Systems Journal* 2, no. 5 (May 1977): 34–35, 38–43. See also Wozniak, *iWoz*, 186–206; and Isaacson, *Steve Jobs*, 71–85. The Apple company soon provided a television monitor to go with the computer, along with external disk drives. Apple began in 1976 as a partnership between Wozniak, Jobs, and an older Atari engineer, Ronald Wayne. The firm began as a partnership with unlimited liability, and Wayne decided to withdraw after two weeks. Mike Markkula reorganized the firm as a corporation with limited liability for the founders.

27. For the price of the Apple II, see *Byte* 2, no. 6 (June 1977): 15. On the VisiCalc application, see Burton Grad, "The Creation and Demise of VisiCalc," *IEEE Annals of the History of Computing* 29,

no. 3 (March 2007): 20–31. Another program, Lotus 1-2-3, took its place in the market. For the expansion of Apple, see *Apple Computer, Inc. The First Decade* (October 1987), 6 pp., in Apple Computer, Inc., Records, 1977–1998, M1007, Series 1, Box 6, Folder 12, Stanford University Library.

28. On Wozniak's plane crash and eventual retirement from Apple, see Wozniak, *iWoz*, 234–267. He continued to serve in a part-time capacity, mainly to represent the firm on occasion.

29. On Jef Raskin, see his obituary in the *New York Times*, February 28, 2005, B11, col. 3. For the Macintosh, see Jef Raskin, "The Genesis and History of the Macintosh Project," February 16, 1981, in Apple Computer, Inc., Records, 1977–1998, M1007, Series 3, Box 10, Folder 3, 5 pp., Stanford University Library.

30. On Xerox PARC, see Tekla S. Perry and Paul Wallich, "Inside the PARC: The 'Information Architects,'" *IEEE Spectrum* 22, no. 10 (October 1985): 62–75; and Michael A. Hiltzik, *Dealers of Lightning: Xerox PARC and the Dawn of the Computer Age* (New York: Harper, 2000). See also Alan C. Kay, "Microelectronics and the Personal Computer," *Scientific American* 237, no. 3 (September 1977): 231–244; and Susan B. Barnes, "Alan Kay: Transforming the Computer into a Communication Medium," *IEEE Annals of the History of Computing* 29, no. 2 (February 2007): 18–30. For initial difficulties getting the graphical interface to work, see Campbell-Kelly, *From Airline Reservations to Sonic the Hedgehog*, 246–251.

31. On the Apple Lisa, see Ceruzzi, *A History of Modern Computing*, 273. Apple also failed to find a market for the Apple III, a computer designed by committee and aimed at a business market. Wozniak did not have a role in the design of either machine.

32. For a technical description of the Macintosh computer, see Gregg Williams, "The Apple Macintosh Computer," *Byte* 9, no. 2 (February 1984): 30–54. See also Cary Lu, *Mac: The Apple Macintosh Book* (Bellevue, WA: Microsoft Press, 1984). For the recollections of those involved, see Andy Hertzfeld, *Revolution in the Valley: The Insanely Great Story of How the Mac Was Made* (Sebastopol, CA: O'Reilly Media, 2005), with a foreword by Steve Wozniak.

33. On the market for the Macintosh computer, see Ceruzzi, *A History of Modern Computing*, 273–276. On the IBM Personal Computer, see Gregg Williams, "A Closer Look at the IBM Personal Computer," *Byte* 7, no. 1 (January 1982): 36–64, 68. For the departure of Jobs from Apple in 1985, see Isaacson, *Steve Jobs*, 180–217.

34. For the development of the IBM Personal Computer, see Campbell-Kelly and Aspray, *Computer*, 225–229. Mark Dean (1957–) grew up in an African American family in the eastern Tennessee Valley. Dean showed an early ability in mathematics and attended the University of Tennessee on a scholarship. After graduation, he joined IBM and made himself an expert in the design of graphics capabilities, which earned him a place on the PC team. See Alan S. Brown, "Mark Dean: From PCs to Gigahertz Chips," *The Bent of Tau Beta Pi* (spring 2015), 22–26. Dean served as lead architect of the IBM PC from 1982 to 1989. The first lead architect for the PC was Lewis Eggebrecht, with David Bradley as lead software engineer. Dean designed the ISA bus with Dennis Moeller. Dean later designed the first microchip to achieve a processing speed in the gigahertz range. Email from Dr. Mark E. Dean to David P. Billington, Jr., September 16, 2019. On the rise of IBM-compatible machines, see Ceruzzi, *A History of Modern Computing*, 277–280.

35. For Bill Gates, see David Allison, "Interview with Mr. William 'Bill' Gates," Computer History Collection, National Museum of American History, Smithsonian Institution, Washington, DC, 1993. For Paul Allen, see Paul Allen, *Idea Man: A Memoir by the Cofounder of Microsoft* (New York: Portfolio/Penguin Books, 2011).

36. For the founding of Microsoft (the original name Micro-Soft changed in 1978) and the IBM contract, see Ceruzzi, *A History of Modern Computing*, 232–236, 269–271. The other Harvard student was Monte Davidoff.

37. On the Windows operating system, see Campbell-Kelly and Aspray, *Computer*, 231–232, 243–247. On the Microsoft-Apple litigation, see *Apple Computer, Inc. v. Microsoft Corp.*, *Federal Reporter*, 3rd Series, vol. 35 (1994), 1435.

38. For the development of software for personal computers, see Campbell-Kelly, *From Airline Reservations to Sonic the Hedgehog*, 201–266. For the dominance of Microsoft's applications in the 1990s, see ibid., 251–259; Ceruzzi, *History of Computing*, 309–313.

39. On the Advanced Research Projects Agency (now DARPA), see Richard Van Atta et al., *DARPA: 50 Years of Bridging the Gap* (Arlington, VA: Defense Advanced Research Projects Agency, 2008). The Information Processing Techniques Office (IPTO) within ARPA funded computing research. See Mitch Waldrop, "DARPA and the Internet Revolution," in ibid., 78–85.

40. For time-sharing on computers, see Campbell-Kelly and Aspray, *Computer*, 186–189.

41. For ARPA's vision of shared access over long distances, see Campbell-Kelly and Aspray, *Computer*, 189–193. See also Janet Abbate, *Inventing the Internet* (Cambridge, MA: MIT Press, 1999). Licklider was a social scientist with an interest in psychology. For his ideas, see J. C. R. Licklider, "Man-Machine Symbiosis," *IRE Transactions on Human Factors in Electronics*, HFE-1, no. 1 (March 1960): 4–11; and Chigusa Ishikawa Kita, "J. C. R. Licklider's Vision for the IPTO," *IEEE Annals of the History of Computing* 25, no. 3 (July–September 2003): 62–77. The two subsequent directors of IPTO, Ivan Sutherland and Robert Taylor, also played important roles. Sutherland was a pioneer in computer graphics and Taylor organized the Arpanet, with Lawrence Roberts of MIT as the first manager of the network. Taylor then headed computer research at Xerox PARC in the 1970s.

42. For Baran's work, see Paul Baran, "On Distributed Communications Networks," *IEEE Transactions on Communications Systems* 12, no. 1 (March 1964): 1–9; and "The Beginnings of Packet-Switching: Some Underlying Concepts," *IEEE Communications Magazine* 40, no. 7 (July 2002): 42–48. For the thinking of Donald Davies, see Donald W. Davies and Derek L. A. Barber, *Communication Networks for Computers* (New York: John Wiley and Sons, 1973). For Baran and Davies and their influence on ARPA, see Abbate, *Inventing the Internet*, 7–41.

43. On the construction of the Arpanet, see Abbate, *Inventing the Internet*, 43–81. For Wesley Clark and the use of interface message processors, see ibid., 51–53. For Robert Kahn and Vinton Cerf and TCP/IP, see ibid., 113–133. See also Vinton G. Cerf and Robert E. Kahn, "A Protocol for Packet Network Intercommunication," *IEEE Transactions on Communications* 22, no. 5 (May 1974): 637–648. The 1974 article described only a Transmission Control Protocol (TCP); Cerf and Kahn later grouped certain functions of the TCP into a separate Internet Protocol (IP), to create TCP/IP.

44. On the uses and later development of the Arpanet, its division into two networks, and privatization of the latter to form the present-day Internet, see Abbate, *Inventing the Internet*, 83–111, 134–145, 181–199. See also Janet Abbate, "Privatizing the Internet: Competing Visions and Chaotic Events, 1987–1995," *IEEE Annals of the History of Computing* 32, no. 1 (2010): 10–22.

45. On the opening of the Internet to a wider public, see Abbate, *Inventing the Internet*, 199–220; and Campbell-Kelly and Aspray, *Computer*, 249–253. The largest providers of proprietary content to consumers in the United States in the 1990s were America Online and CompuServe.

46. On Engelbart, see Susan B. Barnes, "Douglas Carl Engelbart: Developing the Underlying Concepts for Contemporary Computing," *IEEE Annals of the History of Computing* 19, no. 3 (March 1997): 16–26; and Ceruzzi, *A History of Modern Computing*, 259–261. Also see John Markoff, *What the Doormouse Said: How the Sixties Counterculture Shaped the Personal Computer Industry* (New York: Penguin Books, 2005), for a wider cultural history of the San Francisco Bay area in the 1960s and 1970s, the area in which Jobs, Wozniak, and Xerox PARC made their contributions.

47. For Sir Timothy Berners-Lee (1955–), see Tim Berners-Lee with Mark Fischetti, *Weaving the Web: The Original Design and Ultimate Destiny of the World Wide Web* (New York: HarperCollins, 1999). For his insight about the need for simplicity and a solution that could work with diversity, see ibid., 15–16.

48. For the rise of local networking, see Ceruzzi, *A History of Computing*, 291–295. See also Robert M. Metcalfe, "How Ethernet Was Invented," *IEEE Annals of the History of Computing* 16, no. 4 (1994): 81–88.

49. For his original proposal, see the appendix in Berners-Lee, *Weaving the Web*, 211–229. See also ibid., 17–20. For an introduction to early Hypertext Markup Language (HTML), the tags used on web pages, see Elizabeth Castro, *HTML for the World Wide Web* (Berkeley, CA: Peachpit Press, 1996).

50. Nicola Pellow, an intern with CERN, wrote the initial version of the browser that could work with computers other than the NeXT machines that Berners-Lee used to develop the original Web at CERN. Her version had a command-line interface; a version with a more graphical interface eventually took its place. See Berners-Lee, *Weaving the Web*, 29–30, 32–33, 48, 58.

51. For the launching of the Web, see Berners-Lee, *Weaving the Web*, 23–51; and T. Berners-Lee, "WWW: Past, Present, and Future," *IEEE Computer* 29, no. 10 (October 1996): 69–77. For the work of Marc Andreessen, see Ceruzzi, *A History of Modern Computing*, 300–304. Netscape Navigator soon lost market share to Microsoft's Internet Explorer web browser and reemerged as a new web browser, Mozilla Firefox. Other browsers also entered the market in the early twenty-first century, such as Google's Chrome.

52. See Berners-Lee, *Weaving the Web*, 22–23.

53. On the later career of Steve Jobs, see Isaacson, *Steve Jobs*, 211–572.

54. For these quotes, see Isaacson, *Steve Jobs*, 567.

55. See Ralph K. Cavin III, Paolo Lugli, and Victor V. Zhirnov, "Science and Engineering beyond Moore's Law," Special Centennial Issue, *Proceedings of the IEEE* 100 (2012): 1720–1749. On quantum computing, see *Quantum Computing: Problems and Prospects*, ed. Emily Grumbling and Mark Horowitz (Washington, DC: National Academies Press, 2018).

Conclusion

1. See David P. Billington, "Structures and Machines: The Two Sides of Technology," *Soundings: An Interdisciplinary Journal*, no. 3 (fall 1974): 275–288. Quote is from ibid., 278.

2. The transistor was an unusual case, in which the federal government regulated the Bell System in exchange for granting it a monopoly. Reduced competition made it easier for Bell Labs to do basic research. But like the innovators of the microchip a decade later, the Bell System still had a commercial incentive to develop the transistor and then needed military purchases afterward to bring down the cost and develop a civilian market.

3. For the United States, see David Mowery and Nathan Rosenberg, "The U.S. National Innovation System," in *National Innovation Systems: A Comparative Analysis*, ed. Richard R. Nelson (New York: Oxford University Press, 1993), 29–75.

4. For a recent call on behalf of federally supported basic research, see *Restoring the Foundation: The Vital Role of Research in Preserving the American Dream* (Cambridge, MA: American Academy of Arts and Sciences, (2014). The report urges a more flexible and long-range approach.

5. The federal Defense Advanced Research Projects Agency (DARPA) developed an institutional model to generate a flow of deeper innovations by maximizing the intellectual independence of specialists recruited to achieve them. But the DARPA model still looks to advances that can be achieved in 3–5 years. See again Richard Van Atta et al., *DARPA: 50 Years of Bridging the Gap* (Arlington, VA: Defense Advanced Research Projects Agency, 2008).

6. See *The Competitive Status of the U.S. Auto Industry: A Study of the Influences of Technology in Determining International Industrial Competitive Advantage* (Washington, DC: National Academy Press, 1982), 35–50, on the evolution from radical to incremental innovation from the early 1900s to the 1970s. For data on motor vehicle production from 1960 to 2009, see *National Transportation Statistics 2017* (Washington, DC: U.S. Department of Transportation/Bureau of Transportation, 2017), 42. Motor vehicles per person have been roughly stable.

7. See Merritt Roe Smith, *Harpers Ferry Armory and the New Technology: The Challenge of Change* (Ithaca, NY: Cornell University Press, 1977).

8. For overviews of medicine and agriculture, see Joseph Bronzino, Vincent H. Smith, and Maurice L. Wade, *Medical Technology and Society: An Interdisciplinary Perspective* (Cambridge, MA: MIT Press, 1990); and Paul K. Conkin, *A Revolution Down on the Farm: The Transformation of American Agriculture since 1929* (Lexington: University Press of Kentucky, 2008).

9. See Robert J. Gordon, *The Rise and Fall of American Growth: The U.S. Standard of Living since the Civil War* (Princeton, NJ: Princeton University Press, 2016).

10. See Frank Levy and Richard J. Murnane, *The New Division of Labor: How Computers Are Creating the Next Job Market* (Princeton, NJ: Princeton University Press, 2005). For a more pessimistic view, see Erik Brynjolfsson and Andrew McAfee, *The Second Machine Age* (New York: W. W. Norton, 2014). For a more hopeful view of the future, see Derek Thompson, "A World Without Work," *The Atlantic* 316, no. 1 (July–August 2015), 51–61.

11. For an approach to include radical innovators and their works in undergraduate education, see David P. Billington, "Engineering in the Modern World: A Freshman Course in Engineering," 1993 Frontiers in Engineering Conference, Washington, DC, November 1993. See also David P. Billington, *The Innovators: The Engineering Pioneers Who Made America Modern* (New York: John Wiley and Sons, 1996), 1–20. To integrate engineering ideas into secondary school instruction, see David P. Billington, Jr., "Engineering in the Modern World," *World History Bulletin* 24, no. 2 (fall 2008), 22–24.

Index